"Dr. Gans has analysed a very interesting series of recent special operations conducted by a range of countries with varied military cultures. His analysis suggests that special operations continue to have genuinely strategic impact if used to meet strategic ends, and if commanded and controlled at the highest level where policy makers are able to tolerate risk. Gans' analysis also suggests that with the changing character of warfare special forces are more likely to be successful if they are integrated effectively with other capabilities like information and cyber. It is a thoughtful book which should be read by those who want to understand the enduring relevance of special operations and their value in the modern world."

General Sir Nick Carter, *Former UK CDS and Distinguished Visiting Fellow at the Hoover Institute*

"Dr. Gans' fresh look at the value of special operations forces as an integrator when both regular and irregular multidomain capabilities are operating is unique. Commanders and leaders of combined, joint, interagency and multidomain organizations and operations that desire to increase their effectiveness will be well served to investigate the various case studies and insights provided by Dr. Gans in "*Special Operations and National Security: Policies, Strategies, and Tactics*.""

Eric P. Wendt, *Lieutenant General, United States Army, Special Forces (Retired)*
(This endorsement is Lieutenant General (Retired) Wendt's personal view and does not imply DoD, U.S. Army, or U.S. Army Special Forces endorsement.)

"Many will think of Special Operations as helicopter raids conducted by heavily armed soldiers on compounds in far off dusty lands but in reality, they are so much more. This book cleverly introduces the reader to the broad range of capabilities that can be deployed under the banner of a Special Operation. It also shines a light on the risks, threats and opportunities that this type of operation presents to the governments of today. There is plenty here for the military academic and defence college student, but this is also a fascinating and entertaining collection of case studies which goes a long way to showing the versatility, impact and challenges associated with this very modern form of warfare."

Brigadier (Retired) Rob Stephenson, *Former Deputy Commander of the NATO Special Operations Headquarters. He was in the British Army for 36 years and served on operations in Northern Ireland, the Balkans, North Africa, Iraq, and Afghanistan with both the Parachute Regiment and the Special Air Service Regiment.*

"Ben Gans brilliantly illuminates and extends our concept of "special operations" units beyond elite raiders to include cyber forces, proxies and quasi-state militias – which are already fully deployed in global strategic competition. How China, Russia and Iran are aggressively expanding their dominance, often with violence carefully executed below the threshold of declared war, is revealed in present-day cases exceptional for the author's depth of research and penetrating analysis, and their immediate professional value."

Kalev I. Sepp, *PhD, Former Deputy Assistant Secretary of Defense for Special Operations and Counter-Terrorism*

"Ben Gans' volume on special operations is a well-written, meticulously researched book that is a must-read for anyone who is interested in special operations/special forces. His skillful examination of six case studies that capture the varied and dynamic nature of special operations is as entertaining as it is insightful. Gans' publication is a great addition to the body of knowledge and study of special operations."

Colonel (Retired) Bernd Horn, *PhD, Former Deputy Commander Canadian Special Operations Forces Command*

"Gans updates and convincingly argues for the continuing relevance of the dominant theory of special operations–relative superiority. He expands its scope considerably in pursuit of the idea of special operations as an integrating function across various activities and different operating domains simultaneously. His theory, as well as the contemporary case studies contained within, make Gans his work required reading for special operations and national security professionals everywhere."

Dr. James Kiras, *Professor of Strategy and Security Studies, School of Advanced Air & Space Studies*

"In *Special Operations and National Security: Policies, Strategies, and Tactics*, Ben Gans combines deep academic research with a seasoned practitioner's touch to shed light on the often shadowy world of special operations. Through a varied series of case studies, Gans explores what special operations can and cannot accomplish and how they fit into the wider tool kit available to foreign policy decisionmakers. In doing so, *Special Operations and National Security* provides an important contribution to defense analysts exploring this field as well to operators wanting to better understand their profession."

Dr. Raphael S. Cohen, *Senior Political Scientist and Director of the Strategy and Doctrine Program of RAND's Project AIR FORCE*

"In the complex world of modern warfare, it's crucial to bridge theoretical knowledge with practical application. *Special Operations and National Security: Policies, Strategies, and Tactics* endeavors to do just that, offering a comprehensive look at the strategic nuances of special operations in the service of national security. This book is an earnest attempt to enrich the academic and operational discourse surrounding this complex area."

Dr. Ulrica Pettersson, *Director of the Centre of Special Operations Research at the Swedish Defence University*

"Ben Gans delves into the complexities of contemporary special operations in this thought-provoking analysis. Illuminating the nexus between the current strategic environment and the evolving role of special operations, this book provides a novel perspective, positioning special operations as an indispensable integrating force providing strategic effects. This volume serves as essential reading for scholars and practitioners seeking a deeper understanding of this vital instrument of foreign policy."

Patricia J. Blocksome, *PhD, Vice President for Research, Special Operations Research Association*

"Special Operations and National Security offers a compelling explanation for the increasing importance of special operations, how it is evolving in different domains, and future implications. The case studies are illuminating and many contemporary, illustrating the speed of disruption in the irregular threats space. Ben Gans has the practitioner chops and a mastery of the academic aspects of the topic to provide the type of insights policy makers and militaries need to craft effective strategies including SOF in an increasingly dangerous world."

Dr. Craig Whiteside, *Co-Author of* The ISIS Reader *and Professor, U.S. Naval War College*

"A comprehensive tour d'horizon of modern special operations that invites to think beyond traditional notions of such operations and the forces that conduct them. Food for thought for practitioners, policy makers, scholars and all others interested in the evolution of special operations in today's challenging security environment."

Dr. Martijn Kitzen, *Chair Irregular Warfare and Special Operations, Faculty of Military Sciences, Netherlands Defence Academy*

Special Operations and National Security

This book focuses on strategic special operations and how these have led to the achievement of major foreign policy goals, which is illustrated by six case studies.

The study specifically focuses on the alignment of the policies, strategies, and tactics that dominated these operations, providing a fresh perspective. Theoretically, the work underscores the continued relevance of relative superiority as the dominant theory in the field. Importantly, it aligns the potential ways of achieving and sustaining relative superiority with the ability to conduct contemporary special operations across multiple domains, thereby generating cross-dimensional effects. In terms of methodology, the book includes a description and analysis of six distinct case studies in special operations, ranging from short-term surgical strikes to sustained special warfare campaigns and spanning multiple geographic regions – including in Iran, Lebanon, Syria, Ukraine, and the South China Sea. From that perspective, relative superiority theory is extrapolated to explain how this critical condition can be achieved and sustained during protracted special operations. The unique value of this research is underscored by the author's collection of exclusive primary source data on the operations in the respective countries. The book's conclusions explain how the personnel involved fulfill the strategic promise of special operations, given their position within the larger framework of foreign policy.

This book will be of much interest to students of special operations, military and strategic studies, defense studies, and security studies in general.

Ben Gans is a Visiting Professor at the Defense Analysis Department, US Naval Postgraduate School.

Cass Military Studies

Civil-Military Cooperation in International Interventions
The Role of Soldiers
Agata Mazurkiewicz

Contemporary Military Reserves
Between the Civilian and Military Worlds
Edited by Eyal Ben-Ari and Vincent Connelly

Military Strategies of the New European Allies
A Comparative Study
Håkan Edström and Jacob Westberg

Proxy War in Yemen
Bernd Kaussler and Keith A. Grant

Understanding Battlefield Coalitions
Edited by Rosella Cappella Zielinski and Ryan Grauer

Supreme Leadership in Modern War
Civil-Military Relations During Competition and War
Edited by James Lacey and Williamson Murray

The Air War in Ukraine
The First Year
Edited by Dag Henriksen and Justin Bronk

Special Operations and National Security
Policies, Strategies, and Tactics
Ben Gans

For more information about this series, please visit: https://www.routledge.com/Cass-Military-Studies/book-series/CMS

Special Operations and National Security

Policies, Strategies, and Tactics

Ben Gans

Routledge
Taylor & Francis Group
LONDON AND NEW YORK

First published 2025
by Routledge
4 Park Square, Milton Park, Abingdon, Oxon OX14 4RN

and by Routledge
605 Third Avenue, New York, NY 10158

Routledge is an imprint of the Taylor & Francis Group, an informa business

British Library Cataloguing-in-Publication Data
A catalogue record for this book is available from the British Library

Library of Congress Cataloging-in-Publication Data
Names: Gans, Ben, 1981- author.
Title: Special operations and national security : policies, strategies, and tactics / Ben Gans.
Description: Abingdon, Oxon ; New York, NY : Routledge, 2025. |
Series: Cass military studies | Includes bibliographical references and index. |
Summary: "This book focuses on strategic special operations and how these have led to the achievement of major foreign policy goals, which is illustrated by six case studies. The study specifically focuses on the alignment of the policies, strategies, and tactics that dominated these operations, providing a fresh perspective. Theoretically, the work underscores the continued relevance of relative superiority as the dominant theory in the field. Importantly, it aligns the potential ways of achieving and sustaining relative superiority with the ability to conduct contemporary special operations across multiple domains, thereby generating cross-dimensional effects. In terms of methodology, the book includes a description and analysis of six distinct case studies in special operations, ranging from short-term surgical strikes to sustained special warfare campaigns and spanning multiple geographic regions - including in Iran, Lebanon, Syria, Ukraine, and the South China Sea. From that perspective, relative superiority theory is extrapolated to explain how this critical condition can be achieved and sustained during protracted special operations. The unique value of this research is underscored by the author's collection of exclusive primary source data on the operations in the respective countries. The book's conclusions explain how the personnel involved fulfill the strategic promise of special operations, given their position within the larger framework of foreign policy. This book will be of much interest to students of special operations, military and strategic studies, defence studies and security studies in general"-- Provided by publisher.
Identifiers: LCCN 2024028416 | ISBN 9781032429694 (hardback) |
ISBN 9781032429700 (paperback) | ISBN 9781003365112 (ebook)
Subjects: LCSH: Special forces (Military science)--Case studies. |
Special operations (Military science)--Case studies. | Irregular warfare--Europe, Eastern--Case studies. | Irregular warfare--Middle East--Case studies. |
Irregular warfare--China--Case studies.
Classification: LCC U262 .G358 2025 | DDC 356/.16--dc23/eng/20240712
LC record available at https://lccn.loc.gov/2024028416

ISBN: 978-1-032-42969-4 (hbk)
ISBN: 978-1-032-42970-0 (pbk)
ISBN: 978-1-003-36511-2 (ebk)

DOI: 10.4324/9781003365112

Typeset in Times New Roman
by KnowledgeWorks Global Ltd.

Contents

Figures

Foreword

Soon after coming to power in 1985, Soviet leader Mikhail Gorbachev predicted an end to the Cold War followed by a "coming century of peace." While the Cold War *did* end a few years later, peace most assuredly did not diffuse throughout the world. Indeed, the new century has proved rife with wars, one-time uses of pre-emptive or preventive force, and a wide range of strong-arm tactics. The much-respected Geneva Academy of International Humanitarian Law has focused closely on armed conflicts, finding that during the first half of the decade of the 2020s, the number of them ongoing around the world averaged, at any given moment, more than *one hundred.* The vast majority of the violence underway during these years was small-scale, while the Russo-Ukrainian and Israeli-Gazan conflicts, along with civil wars in Sudan and Myanmar, reflect the continuing presence of death-dealing at the higher levels of intensity. In short, Gorbachev's prediction was far off the mark.[1] A more accurate view was provided, back in the 1950s, by a leading scholar of international relations, Kenneth Waltz. As early as 1959, he foresaw that nuclear deterrence between major powers would foster a world where "mutual fear of big weapons may produce, instead of peace, a spate of smaller wars."[2]

Another important thinker of the time, the economist Kenneth Boulding, picked up on this point about a "spate of smaller wars" a few years later, observing that, whatever a nation's inherent capabilities, its ability to intervene far from its own locus of power would diminish with distance. As he put it, this was "the great principle of *the further the weaker* ... This principle is of enormous importance in international relations."[3] Boulding's "loss-of-strength gradient" meant, in practical terms, that the amount of initial capability deployable quickly is necessarily limited – like the eleven Green Beret "A-teams" sent to Afghanistan in the autumn of 2001 – and over time the economic, political, and social costs of maintaining a build-up would prove too onerous to sustain (also as in Afghanistan over two decades).

Waltz and Boulding correctly saw that the global landscape was going to be permeated by all sorts of armed conflicts, many if not most in remote, perilous settings where the "getting to" would be difficult and the ability to act in the way that American soldier-statesman Colin Powell preferred, his idea of "overwhelming force," would prove at best extremely limited. In the world Waltz and Boulding correctly predicted would come to be, leading powers *were* impelled to limit the intensity of wars, and the ineluctable factors of time, space, and sustainability put the onus of action increasingly on special operations forces.

Given that special operators can engage swiftly, at great distances, and across a range of operational settings, political leaders have found them a strategic tool of considerable value. But the fundamental dynamic about their use is that they will almost always face far larger numbers of adversaries throughout their periods of employment – whether in short-duration actions or in more protracted campaigns. This basic truth led one of the leading special operators of the modern era, Admiral William McRaven, to advance a theory of "relative superiority" that could serve to guide raids, rescues, and other short-duration special operations. The basic idea is that elite forces should begin and complete their missions within the window of time provided by tactical, technical, or doctrinal surprise before enemies have had time to recover and gather up sufficient forces to defeat the raiders.[4]

Interestingly, McRaven's insight about relative superiority was not the first expression of this concept. Just a year after Boulding introduced the loss-of-strength gradient, revolutionary leader Che Guevara was writing about relative superiority, both in tactical and longer-range strategic terms. As he put it, "Facing the general superiority of the enemy at a given place, one must find a form of tactics … that unbalances the correlation of forces. In these conditions, tactical victory is assured; if relative superiority is not clear, it is better not to act."[5]

With Che's views in mind, it grows clear that the realm of special operations extends not only to the matter of responding to emergent crises caused by others but also to the choices that are available to those who might wish to instigate all manner of troubles in the first place. The late Qasem Soleimani comes to mind, an Iranian special operator who, with slender but special resources, increased Tehran's influence across an arc of the Middle East ranging from Yemen and Iraq to Syria and Lebanon. Indeed, most of the "internal wars" that populate the Geneva Academy's database reflect Che-like efforts to leverage small units of action with the aim of "unbalancing" the overall correlation of forces to achieve an enduring form of relative superiority.

A world replete with armed conflicts, in which those who would intervene to tamp them down are limited by constraints of time, distance, and sustainability, can only look to special operations forces to have hope of engaging – both quickly and decisively when possible, but for protracted periods when necessary. Colonel Ben Gans realizes this and provides in this volume a thorough, thoughtful exposition of the ways in which special operations have been and are continuing to be employed. The value of doing so is that he is then able to draw important lessons from various examples of recent use. Doing so, he may also be providing the kind of glimpse of the future that Waltz, Boulding, and Che Guevara were able to sketch out when they were writing in that fertile (and critical) period between 1959 and 1963.

That earlier time was a period when nuclear proliferation and the threat of atomic war rose to the fore, and when irregular warfare – for the most part in forms of insurgency and terrorism – was growing increasingly prominent across a global spectrum of conflict. And today we can see that both these concerns still matter very much and require the use of special operations to cope with them. Regarding efforts either to slow or stop a rogue nation's march toward possession of nuclear weapons, Colonel Gans provides two detailed studies. One reflects the operational complexity associated with physically destroying an illicit Syrian proliferation site; the other describes the use of a computer worm to disable centrifuges at an Iranian nuclear facility. This virtual tool of special operations suggests that, in addition to traditional notions of sabotage and preventive raiding, a "cybotage" phenomenon is going to be on the rise.

Gans also turns with much insight into the ways in which special operations have become increasingly attractive to those who would behave aggressively. In this regard, his examination of the skillful, near-bloodless Russian take-over of Crimea by "little green men" offers a primer in such methods – as do his accounts of the expansion of Iranian influence in Lebanon and Chinese reliance on irregular maritime militias to assert control in disputed seas. In all these instances, his very detailed accounts should serve usefully the cause of those who would detect and strive to disrupt such acts of aggression.

Only one of the cases Gans examines resulted in an outright failure: the Russian attempt to seize, hold, and exploit Hostomel Airport during its initial offensive against Kyiv in 2022. But this debacle was nested in the larger Russian inability to break through the swarming defense of small Ukrainian units with their main combat forces. Beyond this example of a special operation gone awry, Gans points out, perhaps implicitly in all his other cases, how such undertakings, when well-planned and integrated with higher policy and strategy, are likely to achieve high rates of success. It is a most sobering thought.

In addition to the insights derived from the cases, Gans has also performed the service of describing ways in which relative superiority can apply to protracted special operations. Where Che Guevara's view of relative superiority as a means of "unbalancing" the correlation of forces was shaped by experience in the Cuban Revolution, Gans has developed this notion across several of his cases. This should lead those who formulate protracted special operations in the future to take note that relative superiority is a concept applicable in this realm – not just short-duration raids. In providing this insight, Gans also reminds us of the complex, multi-dimensional nature of special operations and the value of considering them typologically. For it matters whether an operation is quick and one-time or lasts for months or years. Just as it matters whether it is undertaken independently or in conjunction with main forces. Each type has nuances that require care in crafting.

In his classic *Small Wars,* Colonel Charles Calwell argued the case that "irregular warfare must generally be carried out on a method totally different from the stereotyped system [of conventional operations] … The art of war, as generally understood, must be modified."[6] Colonel Ben Gans' current study is replete with fresh insights into the salience of irregular warfare methods today. Indeed, he has given sufficient guidance for purposes, to paraphrase Callwell, of "modifying the art of war with methods totally different from the stereotyped system." May those special operators who seek ways to counter malign and aggressive uses of irregular methods draw inspiration from this exceptionally thoughtful study.

John Arquilla *Monterey, Spring 2024*

Notes

1 See Mikhail S. Gorbachev, *The Coming Century of Peace* (New York: Richardson & Steirman, 1986); and "Armed Conflicts," accessible at: https://geneva-academy.ch/galleries/today-s-armed-conflicts.
2 Kenneth N. Waltz, *Man, the State, and War: A Theoretical Analysis* (New York: Columbia University Press, 1959), 236.
3 Kenneth E. Boulding, *Conflict and Defense: A General Theory* (New York: Harper & Row, Publishers, 1962), 79. Emphasis in the original.
4 William H. McRaven, *SPEC OPS: Case Studies in Special Operations Warfare: Theory and Practice* (New York: Ballantine Books, 1996).
5 Che Guevara, "Guerrilla Warfare: A Method," *Cuba Socialista* (September 1963), 1–17.
6 Colonel C. E. Callwell, *Small Wars: Their Principles and Practice* (London: His Majesty's Stationery Office, 1906), third edition, 23. Calwell later reached the rank of Major-General.

Bibliography

Armed Conflicts," accessible at: https://geneva-academy.ch/galleries/today-s-armed-conflicts (2024).
Boulding, Kenneth E. *Conflict and Defense: A General Theory* (New York: Harper & Row, Publishers, 1962).
Callwell, C.E. *Small Wars: Their Principles and Practice* (London: His Majesty's Stationery Office, 1906).
Gorbachev, Mikhail S. *The Coming Century of Peace* (New York: Richardson & Steirman, 1986).
Guevara, Che, "Guerrilla Warfare: A Method," *Cuba Socialista* (September 1963).
McRaven, William H. *SPEC OPS: Case Studies in Special Operations Warfare: Theory and Practice* (New York: Ballantine Books, 1996).
Waltz, Kenneth N. *Man, the State, and War: A Theoretical Analysis* (New York: Columbia University Press, 1959).

Acknowledgments

With profound gratitude, I reflect on the privilege of writing this book on special operations at the US Naval Postgraduate School in Monterey, California. The Defense Analysis Department, a beacon of intellectual prowess for selected US and international military professionals, and interagency personnel involved in the USSOCOM-sponsored Special Operations and Irregular Warfare program, has not just been a place of work but my intellectual home. The daily, stimulating discussions on today's complex and dynamic strategic environment and its implications for special operations with the department's distinguished faculty members and exceptional students have been a source of inspiration. I am particularly grateful to Vice Admiral (Ret.) Ann Rondeau, Ed.D., Dr. Scott Gartner, Dr. Carter Malkasian, Colonel (Ret.) Brian Greenshields, and Colonel Shaw Pick, among others, for having me as a visiting faculty member at the NPS. I am deeply indebted to Dr. John Arquilla, whose generous support and countless hours of in-depth discussions have significantly improved the quality of this manuscript.

My journey to write this book would not have been possible without the unwavering support of my employer, the Royal Netherlands Army. I am grateful to Lieutenant General Jan Swillens and Brigadier General Rejanne Eimers for their trust in my abilities. I am acutely aware that my growth as a military officer is a testament to the support and camaraderie of my closest colleagues, with whom I have had the honor to serve: Sven, Erik, Hans, Maarten, Sander, Joep, Michel, Huub (x2), Gerto, Richard, Mark (x2), Menno, Sil, Sjoerd, Danny, Jeffrey, Bas (x2), and above all, Ron, to whom I owe an immeasurable debt of gratitude. I am particularly thankful to Brigadier General René van den Berg, a man I deeply admire, for his mentorship and leadership over the past two decades.

To me, there's no greater privilege in life than being part of the military, particularly the unit I so proudly served in. I am aware, however, that this unit is, at times, very demanding. Social support offered by a network of volunteers illustrates this unit's unique social cohesion. Therefore, I will donate my profits from this book, a testament to our collective efforts and dedication, to the Commando & Family Foundation, a cause that is close to my heart and supports those in difficult times.

I am deeply grateful for the support of the many people who generously shared their insights on the special operations discussed in this volume. Given these missions' secretive and sensitive nature, many were only willing to share their experiences on a non-attribution basis. I am deeply grateful for your trust! My gratitude goes to Bobby, Duncan, Blake, Robert, Leah, Bill, Michael, Martijn, and Niek. I am very much indebted to those who wished to remain anonymous. Your insights proved invaluable, and I often think back to the numerous insightful conversations we had worldwide. The community knows who you are.

I thank Andrew Humphreys, Jasmine Erice-Harling, and Max Gethings at Taylor and Francis for selecting this book for their Military, Strategic, and Security Studies series. I am also thankful to Alison Scharmota at NPS GWC and Aart Brutel in the Netherlands for their detailed reviews of the manuscript, as I am to Eldar Hagen for his highly professional artwork, which is included in the book.

This work describes contemporary history, and the special operations discussed are surrounded by a high degree of secrecy. So, undoubtedly, gaps exist in the analysis, and more details of the operations will likely be revealed in the future. Therefore, I tried to offer the most precise understanding of these endeavors possible at this time. Any factual errors or misinterpretations are, of course, mine and mine alone.

This entire project would have remained a distant dream without my family's unwavering and steadfast support. Maion, the love of my life, who courageously and selflessly embraced our American adventure, and our two wonderful sons, Valentijn and Maxim, to whom I am eternally grateful. Voor altijd samen!

Acronyms and Abbreviations

A2AD	Anti-Access Aerial Denial
AEOI	Atomic Energy Organization of Iran
AFO	Advanced Force Operations
AGL	Above Ground Level
AI	Artificial Intelligence
AIS	Automatic Identification System
AIVD	Algemene Inlichtingen en Veiligheidsdienst
AO	Area of Operations
APC	Armored Personnel Carrier
APT	Advanced Persistence Threat
ARC	Autonomous Republic of Crimea
ASEAN	Association of Southeast Asian Nations
ASSR	Autonomous Soviet Socialist Republic
BCE	Before the Common Era
BTG	Battalion Tactical Group
C2	Command and Control
CCG	China Coast Guard
CCP	Chinese Communist Party
CCTV	Closed-Circuit Television
CIA	Central Intelligence Agency
CLCS	Commission on the Limits of the Continental Shelf
CNO	Computer Network Operations
COIN	Counterinsurgency
CONOPS	Concept of Operations
CPI	Corruption Perceptions Index
DA	Direct Action
DEA	Drug Enforcement Agency
DIME(FIL)	Diplomatic, Information, Military, and Economic (Financial, Intelligence, and Law enforcement
DLL	Dynamically Linked Libraries
DOD	Department of Defense
DOE	Department of Energy
DPR	Donetsk People's Republic

DRC	Democratic Republic of the Congo
EA	Electronic Attack
ECU	Eurasian Customs Union
EEU	Eurasian Economic Union
EFP	Explosively Formed Penetrator
EMS	Electromagnetic Spectrum
EU	European Union
EW	Electronic Warfare
FBI	Federal Bureau of Investigation
FLEC	Fisheries Law Enforcement Command
FSB	Federal Security Service
GMTI	Ground Moving Target Indication
GRU	Main Intelligence Directorate
HAHO/HALO	High-Altitude High Opening/High-Altitude Low Opening
HRO	Hostage Release Operations
HQ	Headquarters
HUMINT	Human Intelligence
IADS	Integrated Air Defense System
IAEA	International Atomic Energy Agency
IAF	Israeli Air Force
ICBM	Intercontinental Ballistic Missile
ICS	Industrial Control Systems
IDF	Israeli Defense Force
IDS	Intrusion Detection Systems
IE	Information Environment
IJO	Islamic Jihad Organization
IMINT	Imagery Intelligence
IoT	Internet of Things
IPS	Intrusion Prevention Systems
IRGC	Islamic Revolutionary Guard Corps
IS	Information Systems
ISIS	Islamic State of Iraq and Syria
ISR	Intelligence, Surveillance, and Reconnaissance
JADO	Joint All Domain Operations
JFC	Joint Force Commanders
JSOC	Joint Special Operations Command
KGB	Committee for State Security
KSSO	Komandovanie Sil Spetsial'nalnykh Operatsii
LAF	Lebanese Armed Forces
LAN	Local Area Network
LEU	Low-Enriched Uranium
LRPA	Long Range Patrol Aircraft
MAAG	Military Assistance Advisory Group
MANPADS	Man-Portable Air Defense Systems
MDO	Multi-Domain Operations

MEK	Mujahedin-e Khalq
ML	Machine Learning
NATO	North Atlantic Treaty Organization
NCA	National Command Authority
NCRI	National Council of Resistance of Iran
NGA	National Geospatial-Intelligence Agency
NPT	Nonproliferation Treaty
NSA	National Security Agency
NSDC	National Security and Defense Council
NSS	National Security Strategy
OE	Operational Environment
OODA	Observe, Orient, Decide, and Act
OPE	Operational Preparation of the Environment
OPSEC	Operational Security
OS	Operating System
OSS	Office of Strategic Services
PCA	Permanent Court of Arbitration
PCG	Philippine Coast Guard
PLA	People's Liberation Army
PLAN	People's Liberation Army Navy
PLC	Programmable Logic Controller
PLO	Palestine Liberation Organization
PM	Prime Minister
PMF	Popular Mobilization Forces
PRC	People's Republic of China
RAF	Royal Air Force
RAS	Robotic and Autonomous Systems
RMA	Revolution of Military Affairs
RW	Rotary Wing
SAM	Surface-to-Air Missile
SAS	Special Air Service
SBGS	State Border Guard Service
SCADA	Supervisory Control and Data Acquisition
SCS	South China Sea
SFSR	Soviet Federative Socialist Republic
SIGINT	Signals Intelligence
SOE	Special Operations Executive
SOF	Special Operations Forces
SOFA	Status of Forces Agreement
SR	Special Reconnaissance
SSO	Sily Spetsial'nykh Operatsii
TAC	Treaty of Amity and Cooperation
TALO	Tactical Air Landing Operation
TEO	Technical Exploitation Operation

TTPs	Tactics, Techniques, and Procedures
UAE	United Arab Emirates
UAS	Unmanned Aerial Systems
UK	United Kingdom
UN	United Nations
UNCLOS	United Nations Convention on the Law of the Sea
UNSC	United Nations Security Council
US(G)	United States (Government)
USCYBERCOM	U.S. Cyber Command
USSOCOM	United States Special Operations Command
UW	Unconventional Warfare
VDV	Vozdushno-desantnye voyska Rossii
WAAF	Women's Auxiliary Air Force
WMD	Weapons of Mass Destruction
WWI	World War I
WWII	World War II

Part I

Introduction

1 Special Operations

Theory and Strategic Utility

Introduction

On the night of June 16, 1943, a Westland Lysander flew a British wireless radio operator code-named "Madeleine" and two other British operatives to a moonlit field in the Loire Valley, northern France. On the ground, Madeleine was met by Henri Déricourt, a French operative working for the British. Within minutes after landing, the airplane took off again, and Madeleine set out for Paris, where she was to join the "Prosper Network" and perform her task as a covert radio operator. Her mission was a dangerous one, as the Germans had given high priority to detecting, locating, and capturing radio operators. To avoid detection, Madeleine continuously changed her hideout and only made brief transmissions. To do so, she had to carry her wireless set while walking the streets of Paris, which were full of German military and intelligence personnel who randomly stopped people and searched them. Madeleine turned out to be extremely well-qualified for this covert action. So, who was this brave woman who carried her 30-pound transceiver with her in a two-foot-long suitcase while evading German detection? Why was she selected for this daunting mission? And what training had she received to conduct this highly specialized and dangerous task?

Her name was Noor Inayat Khan, and she was the daughter of an Indian-born father and an American mother. Her father descended from the legendary "Tiger of Mysore" Tipu Sultan, one of India's greatest warrior princes. While touring the United States (US) with the Royal Musicians of Hindustan, he met Ora Ray in San Francisco and returned with her to the European continent. Noor was born on January 1, 1914, in Moscow, which was then the capital of Imperial Russia, and grew up in Suresnes, a suburb of Paris. Noor loved art, music, and poetry and was a talented musician, playing the piano and the harp. After she graduated from Sorbonne Université with a degree in child psychology, Noor wanted to write and illustrate children's tales and planned to establish an illustrated newspaper specifically for children named "*Bel Age*" (Figure 1.1).

After the outbreak of World War II (WWII), Noor and most of her family felt compelled to flee France for England. Noor joined the British Women's Auxiliary Air Force (WAAF), an organization that served in support of the Royal Air Force (RAF). She was trained as a radio operator and excelled in demonstrating significant

DOI: 10.4324/9781003365112-2

Figure 1.1 Special Operations Executive (SOE) operative Noor Inayat Khan, code-named "Madeleine" (Courtesy: Shrabani Basu).

skills in communicating through a special type of Morse code, skills Noor mastered easily because of her excellent technique on the piano. On November 10, 1942, Noor's ambitions to become more directly involved in the war effort were met when she was invited for a secret interview, supposedly with the War Office. The invitation said she had essential language skills that might make her useful for operational purposes. She was about to be recruited as a covert radio operator for the Special Operations Executive (SOE), a new secretive organization that was directed by British Prime Minister Winston Churchill "to set Europe ablaze" by conducting reconnaissance, supporting resistance movements, and conducting sabotage against the Axis Powers.[1] While performing these tasks, SOE operatives needed to maintain communications with London. To prevent the Germans from learning the content of these messages, they had to be encrypted via Morse code. Noor was an ideal candidate for this task as she had already been trained as a radio operator and in the use of Morse code. Furthermore, the fact that she was a native French speaker and grew up in France made her an excellent candidate for SOE's French (F-) Section, which worked closely with the French resistance.

Only days after Noor's arrival in Paris, the Prosper Network she had joined was compromised, and many SOE operatives were arrested. Based on the information the Germans had found while making these arrests, many other members of the resistance network were arrested in the following days. Despite multiple offers from London to return to England for her safety, Noor decided to stay in Paris and continue her task, which was critical at this time in the war as the Allied invasion of Normandy was less than one year away. Under these immense pressures, Noor was extraordinarily successful and eluded the Germans for months. Noor's success resulted from her keen intuition, constant movement, and fast coding skills. Her eventual arrest was not the result of German excellence in detecting and locating her, nor tactical missteps from her side, but instead because she was betrayed by a woman referred to as "Renée." After her arrest, the Gestapo took Noor to their notorious headquarters on 84 Avenue Foch in Paris, known as "The Street of Horrors." After a month of brutal interrogation, Noor did not betray other members of SOE or the French Resistance. She was then brought to a high-security prison in Pforzheim in southwestern Germany, where she was further interrogated. On September 11, 1944, Noor was transported to Dachau concentration camp in southern Germany, where she and three other female British SOE operatives were killed the next day. Noor was posthumously awarded the French Croix de Guerre and the British George Cross for her actions. Just outside *Fazal Manzil*, meaning the House of Blessing, Noor's happy childhood home, a plaque is dedicated to her and says:

Ici habitait Noor Inayat Khan
1914–1944
Madeleine dans la résistance
fusillée à Dachau
opératrice radio des réseaux buckmaster
Croix de Guerre 1939–1945 George Cross.

Noor will always be remembered as one of the courageous SOE operatives who sent and received coded messages that contributed to the escape of many Allied airmen, arranged false identity papers for secret agents, and identified the exact locations for airdrops. Furthermore, Noor assisted with obtaining weapons and money for members of the French resistance and communicated the exact hiding spots of German heavy weaponry. In doing so, Noor's actions contributed disproportionally to the war effort and demonstrated skills that expanded the range of options available to policymakers that lay outside the scope of conventional military forces. Her actions and skills are, therefore, justly associated with special operations.

Special operations: definitions, terminology, and concepts

Special operations are unique ways (i.e., tactics and operations) of employing means to achieve strategic ends that are not strictly military. Various actors, including military forces, paramilitary forces, intelligence personnel, and non-state actors, can perform special operations. From a Western perspective, the military forces capable of executing special operations can be divided into conventional military forces and Special Operations Forces (SOF). SOF are considered unique military personnel that expand the range of options provided by conventional military forces and, therefore, are more suitable for conducting special operations.[2] In Western military doctrine, SOF are defined as "small, specially organized units manned by carefully selected people using modified equipment and trained in unconventional applications of tactics against strategic and operational objectives."[3] According to Colin Gray, the ability of these small, specially organized units to achieve strategic objectives lies in their economy of force and expansion of strategic choice.[4] Economy of force refers to the ability of special operations to "achieve significant results with limited forces."[5] Expansion of strategic choice refers to the expansion of "the options available to political and military leaders."[6] These two features, in turn, explain the strategic utility of SOF or other forces capable of conducting special operations.[7] The strategic utility of special operations has its pros and cons. The latter, particularly, regarding the risk of failure, results in a disproportionate, negative return on investment.

In Western military literature, special operations are a young yet emergent field in military history and theory. Differences of opinion, however, exist on the precise definition of special operations and have led to its evolvement over time. In the 1970s, Major General (Ret.) Joseph Lutz defined special operations as "secondary or supporting operations which may be adjunct to various other operations and for which no one service is assigned primary responsibility."[8] For reasons discussed later, this initial definition focused on special operations and did not refer to SOF. The definition's uniqueness, therefore, lies in its focus on the relationships between ways and ends without explicitly considering the means to achieve those ends. By 1982, the US Department of Defense (US DOD) decided upon a formal definition of special operations:

> Military operations conducted by specially trained, equipped, and organized DOD forces against strategic or tactical targets in pursuit of national military, political, economic, or psychological objectives. They may support conventional operations, or they may be prosecuted independently when the use of conventional forces is either inappropriate or infeasible. Sensitive peacetime operations, except for training, are normally authorized by the National Command Authorities (NCA) and conducted under the direction of the NCA or designated commander. Special operations may include unconventional warfare, counterterrorist operations, collective security, psychological operations, and civil affairs measures.[9]

This formal definition highlights three critical elements of special operations. First, the US DOD refers to the specially trained, equipped, and organized US DOD forces. SOF, however, still had not been specifically mentioned. Second, the definition provides more insight into what special operations might entail by introducing five core activities, some of which are still included in today's Western doctrines. Finally, the US DOD notes that besides in times of war, special operations may be conducted in peacetime as well. In doing so, the US DOD recognized the sensitive character of special operations; hence, decision authority lies at the NCA level.

The US is certainly not the only military power that formally described special operations and SOF in its military concepts and doctrine. The People's Republic of China (PRC), for example, refers in its military concepts and doctrine to "campaign special operations," which are defined as:

> Irregular operational activities implemented by specially formed, trained and equipped crack force units (element) using special warfare in order to achieve specific campaign and strategic goals. Its goals are mainly to raid the enemy's vital area targets, paralyze the enemy's operational systems {*tixi*}, reduce the enemy's operational capability, and interfere, delay and disrupt the enemy's operational activities in order to create favorable conditions for the main force units.[10]

Although not explicitly mentioned, the definition emphasizes that a key characteristic of special operations is exploiting the enemy's critical vulnerabilities. To this end, special operations must be planned, prepared, and executed in a concealed state.[11] Furthermore, the Chinese list five basic patterns of special operations campaigns: special reconnaissance, special raid attacks, special sabotage attacks, special raid harassments, and special technical warfare.[12] The first four patterns correspond closely to the principal tasks listed in Western concepts and doctrine, although they are defined differently. However, special technical warfare (e.g., electronic warfare and computer network operations) is in Western militaries typically not associated with special operations but rather with cyberspace operations. Finally, China views these five basic patterns as unorthodox (*chuqi zhisheng*) tactics that should be employed in conjunction with orthodox tactics (i.e., conventional military operations) to gain victory over an opponent.[13] This view on integrating conventional forces and special operations has been a longstanding tradition in Chinese military thinking. It can be found in the strategic thought of Sun Tzu, who argued that: "in all fighting, the direct method may be used for joining battle, but indirect methods will be needed in order to secure victory."[14]

Russia also has a longstanding history of special operations. Classical Soviet doctrine refers to *Spetsalnaya Naznacheniya* (Special Purpose) or *Spetsnaz* (Special Designation) operations.[15] These operations, often sensitive in nature, are employed to significantly weaken the political-military capabilities of an adversary through espionage, sabotage, or subversion.[16] Additionally, these operations may be focused on shaping the environment to enable follow-on military operations.[17] Although *Spetsnaz* forces are unquestionably considered elite forces, they are not always the same as the Western conception of SOF.[18] Indeed, the Russians use the word *Spetsnaz* in various ways, including the type of unit (i.e., combat engineers, infantry, or signalers), the unit's status, or whether the unit conducts sensitive and clandestine activities.[19] Current Russian military documents describe the emergence of a Special Operations Forces Command (*Komandovanie sil spetsial'nalnykh operatsii*, or KSSO) and Russian SOF (or *sily spetsial'nykh operatsii*, or SSO).[20] Specific features of Western SOF greatly inspired the establishment of these units in 2008–2012 and illustrate the creation of twenty-first century Russian special operations and SOF.[21] Today, the Russian Ministry of Defense uses the following definition for a special operation:

> The special operation of troops (forces) is a complex of special actions of troops (forces), coordinated by objectives and tasks, time and place of execution, conducted according to a single concept and plan in order to achieve certain goals. Special actions of troops (forces) are activities carried out by specially designated, organized, trained and equipped forces, which apply methods and ways of fighting not typical for conventional forces (special reconnaissance, sabotage, counterterrorist, counter-sabotage, counterintelligence, guerrilla, counter-guerrilla and other activities).[22]

Moreover, Alexander Sladkov describes the strategic objectives for which Russian special operations are employed as:

> Special Operations Forces (SSO) – troops intended to achieve political and economic goals in any geographical part of the world of interest to the Russian Federation.... They come in cases when diplomatic methods are no longer active. Distracting forces and the attention of certain countries by external problems, problems creating them inside, rocking the political systems of these countries, destabilizing the situation, including through a 'third hand.' Special Operations Forces create, train, and supervise foreign guerrilla movements, eliminate unwanted leaders without any sanctions on foreign soil, and so on... Russian experts' main task is the protection of our citizens abroad, the release of Russians who have fallen hostage somewhere in distant areas, and protecting the interests of our country.[23]

Finally, Aleksey Nikolskiy asserts that Russian SOF are "proper combat units themselves and can operate independently. They are ready for rapid deployment across a spectrum of counterterrorism and combat missions, on Russian territory and abroad."[24]

The sensitive character of special operations described by all three military powers can be explained through the application of security as one of the principles of war, which provides a common direction for planning, preparing, and executing military operations. The principle of security is addressed by Brigadier (Ret.) Maurice Tugwell and David Charters, who add the three levels of secrecy (i.e., overt, covert, and clandestine) related to the planning, preparation, and execution of special operations to their definition:

> Small-scale, clandestine, covert, or overt operations of an unorthodox and frequently high-risk nature, undertaken to achieve significant political or military objectives in support of foreign policy. Special operations are characterized by either simplicity or complexity, by subtlety and imagination, by the discriminate use of violence, and by oversight at the highest level. Military and nonmilitary resources, including intelligence assets, may be used in concert.[25]

Drawing from this definition, special operations may be conducted in an overt, covert, or clandestine manner and could be employed during peace or war. According to Gray, the ability to perform special operations covertly or clandestinely, particularly in peacetime, provides strategic leaders with unique options to achieve strategic outcomes and thus underscores their strategic utility.[26]

Overt special operations do not require specific measures of operational security to conceal the operation, its effects, or the authorities involved. The North Atlantic Treaty Organization (NATO) defines an overt operation as "conducted openly, without concealment."[27] Special operations may be conducted overtly when there are no political or military reasons that necessitate their secrecy or when the authorities also have no reasons to require secrecy. Overt operations may be undertaken in peacetime, requiring permission from a host country, and in most cases, if cooperation is wanted, it will be organized in consultation with that country. As a result, it is not required to safeguard the secrecy of the operation as well as the exposure of the authorities. Overt special operations may also be conducted in times of war. Given the offensive nature of that type of special operation (i.e., ambushes and raids), it is not possible, necessary, or desirable to safeguard the secrecy of the operation or the sponsor involved. In highly exceptional cases, such as hostage release operations (HRO), overt special operations conducted in peacetime can also be offensive.

Special operations are conducted covertly when protecting the authorities involved is vital. NATO defines covert operations as: "planned and conducted so as to conceal the identity or permit plausible deniability of the executor."[28] Special operations are conducted covertly when vital politico-strategic and military-strategic interests and risks are involved, requiring the authorities to remain secret. While it is acceptable for the operation itself, or at least some of its effects, to be detectable, the planning, preparation, and execution phases must be conducted so that the operation and its effects cannot be traced back to the authorities. Thus, covert operations are undertaken below the attribution threshold. This type of special operation may also be conducted in peacetime as well as in war.

Whereas covert operations aim to protect the authorities, clandestine operations are designed to prevent the detection of the activities, as well as the effects. NATO defines clandestine operations as "planned or conducted in such a way as to assure its secrecy or concealment."[29] Special operations are conducted clandestinely when vital politico-strategic or military-strategic interests and risks are involved, requiring the operation and the authorities to remain hidden. Clandestine operations are also performed during peacetime and in war. Importantly, given that clandestine operations are conducted below the detection threshold, the activities involved with these operations have an inherent non-kinetic character.

There is a functional distinction between the three types (i.e., overt, covert, and clandestine) based on the degree of secrecy and the ensuing implications for special operations. Given the complexity of the strategic environment, in which a myriad of actors with diverse interests plays a role and combatants and non-combatants move among each other, it is essential to consider the perspective from which a special operation is seen as overt, covert, or clandestine. For example, a special operation is conducted overtly from the host nation's perspective when the local authorities are aware of both the operation and the sponsor, albeit with compartmentalized knowledge, and they may also be involved in the operation itself. At the same time, operational security measures may be in place to protect both the operation and the sponsor from a state or non-state adversary active in the host nation's territory, given the clandestine character of the special operation, as seen from that perspective.

The two most influential doctrines on special operations today are NATO's Allied Joint Doctrine for Special Operations AJP-3.5(B) and US Joint Publication 3-05 Special Operations.

NATO doctrine defines special operations as:

> Military activities conducted by specially designated, organized, selected, trained, and equipped forces using unconventional techniques and modes of employment. These activities may be conducted across the full range of military operations, to help achieve the desired end-state. Politico-military considerations may require clandestine or covert techniques and the acceptance of a degree of political or military risk not associated with operations by conventional forces. Special operations deliver strategic or operational-level results or are executed where significant political risk exists.[30]

Several observations can be made from this definition. First, NATO refers to a special type of forces that operate unconventionally. Second, NATO distinguishes special operations from conventional operations, aiming at maximum effectiveness in situations with significant politico-strategic or military-strategic interests and risks for which strict confidentiality about the operation and the sponsor is required.

US military doctrine provides the following definition of special operations:

> Special operations require unique modes of employment, tactics, techniques, procedures, and equipment. They are often conducted in hostile, denied, or politically and/or diplomatically sensitive environments, and are characterized by one or more of the following: time-sensitivity, clandestine or covert nature, low visibility, work with or through indigenous forces, greater

> requirements for regional orientation and cultural expertise, and a higher degree of risk. Special operations provide joint force commanders (JFCs) and chiefs of mission with discrete, precise, and scalable options that can be synchronized with activities of other interagency partners to achieve United States Government (USG) objectives.[31]

This definition highlights several features of special operations absent from NATO's. First, the US military doctrine refers to time-sensitivity, implying, among other things, the precise right moment of mission execution, called by the ancient Greeks: "*kairos*" (ὁ καιρός), the propitious moment in war. Second, the definition emphasizes the role of local partner forces that may be involved in special operations and require personnel to have specific knowledge and skills. Finally, the US places special operations within an interagency context, thereby underlining the importance of coordination with other US government agencies.

The common element between these two dominant definitions is that special operations require unconventional modes of employment. This observation is shared by James Kiras, who explicitly points to the constraints involved in the employment of conventional forces by proposing that special operations are:

> Unconventional actions against enemy vulnerabilities in a sustained campaign, undertaken by specially designated units, to enable conventional operations and/or resolve economically politico-military problems at the operational or strategic level that are difficult or impossible to accomplish with conventional forces alone.[32]

By pointing directly to the limitations of conventional forces, Kiras emphasizes the distinctive character of the specially designated forces that can employ unconventional modes of operation. These specially designated forces and unconventional actions should not be considered superior to conventional forces but complementary. This observation is shared by John Arquilla, who demonstrated through various historical examples that integrating the two types of military operations will lead to success.[33]

Admiral (Ret.) William McRaven explains the need for such specially designated forces as critical to overcoming what Carl von Clausewitz described as the friction of war. Clausewitz states that friction is "the concept that differentiates actual war from war on paper."[34] The concept includes various unforeseen events that may occur during the execution of military operations and make even the best plan challenging to execute because these unexpected events are difficult to manage. McRaven argues that SOF could overcome the friction of war by achieving a condition he defined as relative superiority:

> I will show that through the use of certain principles of warfare a special operations force can reduce what Carl von Clausewitz calls the frictions of war to a manageable level. By minimizing these frictions, the special operations force can achieve relative superiority over the enemy. Once relative

> superiority is achieved, the attacking force is no longer at a disadvantage and has the initiative to exploit the enemy's weaknesses and secure victory.[35]

According to McRaven, to achieve relative superiority, six specific principles of special operations must be applied and integrated throughout the planning, preparation, and execution phase of an operation. The principles are simplicity, security, repetition, surprise, speed, and purpose.[36] By applying these principles, SOF can reduce the friction of war to the extent that they can achieve a competitive advantage over an opponent. In doing so, a small special operations force can defeat a numerically superior and well-prepared opponent. Although his unique monograph has significantly contributed to developing a theory of special operations, his focus is on one specific special operations mission set, namely, Direct Action (DA).

Robert Spulak Jr. elaborated on McRaven's work by identifying specific SOF qualities that transcend conventional forces' capabilities.[37] According to Spulak, the interrelation between ways (i.e., special operations) and means (i.e., SOF) to effectively cope with the frictions of war is best illustrated by the argument that:

> Special operations are missions to accomplish strategic objectives where the use of conventional forces would create unacceptable risks due to Clausewitzian friction. Overcoming these risks requires special operations forces that directly address the ultimate sources of friction through qualities that are the result of the distribution of the attributes of SOF personnel.[38]

According to Spulak, the distinctive attributes of SOF, by which they address the sources of friction, include being elite warriors who operate creatively and flexibly.[39] Given the ubiquitous use of Clausewitzian friction, updating the concept seems necessary. Although he considers "*Friktion*" mainly an unpredictable factor [*Zufall*, chance] – he mentions, for instance, the weather – to be overcome by "an iron will," modern warfare obviously still has to do with "chance." However, in contrast to the German iron will of the nineteenth century, today, flexibility and creativity, as indicated by Spulak, are the essential components of successful modern specially selected forces.

Due to the broadness of special operations, Lieutenant General (Ret.) Eric Wendt proposes to divide SOF into two main types: Direct SOF and Force Multiplier SOF.[40] Direct SOF are forces that are trained and equipped to conduct special operations independently.[41] In that regard, Direct SOF are optimized for SR and DA mission sets. Force Multiplier SOF, by contrast, are trained and equipped to "force multiply" and are capable of conducting operations by, with, or through partner forces.[42] Thus, Force Multiplier SOF follow an indirect approach and are, therefore, most suitable for MA. According to Wendt, the distinctive features between these two types of SOF do not indicate that one type is better than the other but rather the distinction in their strategic utility.[43] The strategic utility of these two types of SOF is best illustrated by Wendt's metaphor in which he compares them with a sprinter and a marathon runner who both win an Olympic medal. Wendt points out that both are tremendous athletes, but their utility is tailored to what they intend to achieve.[44]

The conception that special operations require highly specialized people, i.e., specifically trained personnel, raises the question: which came first, a special type of operation or the forces formed with these specialized capabilities? This question is particularly relevant when reflecting on the changing conduct of warfare and its implications for future special operations.

Special operations can only be defined regarding resources, capabilities, or competencies, with an understanding of the type of conflict and mission set in which they perform. Suppose special operations are employed for reconnaissance, sabotage, support to resistance, or to complement conventional forces, to name a few; their strategic impact will depend on their suitability to the specific context.[45] According to Gray, the historical context in which special operations were conducted is so broad that it is impossible to define the unifying conditions and circumstances of their success.[46] He concludes that the strategic utility of special operations depends on their historical context.[47] In other words, it is the changing conduct of warfare, continuously leading to new forms, resulting in the emergence of specific mission sets defined as "special" within their respective context. The sensitivity to a particular context is emphasized by Arquilla, who described special operations as: "that class of military (or paramilitary) actions that fall outside the realm of conventional warfare during their respective periods."[48] These actions may require changes in capabilities, tradecraft, modes of operation, as well as tactics, techniques, and procedures (TTPs). Gray reminds us, however, that despite these changes, the strategic utility of special operations will not change and may remain constant for decades or even centuries to come.[49]

While the strategic utility of special operations is constant, the same cannot be said about the individuals or units that conduct these operations, as requirements will likely change over time. Indeed, the selection, training, and deployment of the brave men and women who have performed special operations throughout history are the determinants that gave meaning to the definition of these specially trained personnel. In retrospect, the sabotage, support to resistance, and reconnaissance missions in occupied Europe and Southeast Asia against the Axis Powers during WWII, stay-behind operations in Western Europe during the Cold War, scud hunting in the Gulf War, and targeting operations in the Afghanistan and Iraq campaigns, among others, are contextual illustrations for how the changing conduct of warfare has led to the definition of special operations and subsequently specially trained personnel within their specific historical context.

The challenge of identifying future special operations and selecting, training, and equipping the appropriate personnel is well-documented in the Office of Strategic Services (OSS) Assessment Staffs' "*Assessment of Men: Selection of Personnel for the Office of Strategic Services*." In short, the OSS training staff was challenged by the fact that:

> It was many months before our conceptions of the different jobs were more than half accurate. We were given the briefest possible time in which to prepare. No one arranged a preliminary world tour for the staff so that the conditions at each base and the operations in progress could be observed at first

> hand. The information that came in from the theaters was scanty; and even if it had been ample and adapted to our purpose, there would have been too much to learn in the time available, too much to remember. Not until much later did some of us who visited installations in the field come to realize the magnitude of the discrepancy between even the better job descriptions—those received in the latter months—and the various dispositions and skills that were actually required in the field.[50]

These observations indicate that as the conduct of warfare changes over time and leads to new forms, understanding why certain mission sets and personnel are defined as special operations and specially trained personnel only becomes possible with hindsight. Consequently, it appears impossible to provide a unified definition of special operations.

Special operations and strategy

The ultimate purpose of special operations is to generate strategic effects supporting political objectives. Special operations are thus a strategic tool that presents unique options for policymakers to defend and advance national interests. To this end, tactical activities are conducted in the strategic environment to exploit enemy vulnerabilities, the consequences of which are known as strategic effects.[51] While policymakers define political objectives, Military Command is responsible for employing tactics.[52] Meeting the achievement of political objectives and the employment of tactics is what Gray defines as strategy.[53] In other words, strategy is about how to apply tactics and operations for political purposes.

Given the clear distinction between the authority and responsibility of the policymakers and the authority and responsibility of the military commanders, exploiting the full potential of special operations is thus highly dependent on bilateral consultation and mutual understanding of strategy development and tactical behavior.[54] Understanding the conversion of tactical performance into strategic effects requires a strategic sense.[55] From a special operations perspective, political objectives will only be secured if they can be achieved through tactical action. Conversely, superior tactical action becomes meaningless if it serves an ill-defined political goal. Fulfilling the strategic promise of special operations requires aligning policies, strategies, and tactics appropriate to the strategic, operational, and tactical levels of war.

The general theory of strategy distinguishes between grand strategy and military strategy. Grand strategy refers to the "direction and use made of any or all the assets of a security community, including its military instrument, for the purposes of policy as decided by politics."[56] Military strategy is more narrowly defined as "the direction and use made of force and the threat of force for political purposes."[57] The challenge for the strategist is thus to unite the employment of military power with political statecraft.

Of the two classical yet distinct schools of thought on military strategy, the first focuses on the strategic thinking of Clausewitz and Antoine-Henri Jomini and their description of a "direct approach." Indeed, Clausewitz emphasizes that destroying an enemy's forces is the overarching principle of war.[58] Destruction requires a direct engagement that involves all available forces.[59] He further emphasizes that the most common component in victory is having superiority in numbers that should be employed in the engagement.[60] This major battle should be considered the center of gravity of an entire military campaign.[61] Jomini also expounded the same principles, something he describes as "offensive actions to mass forces against weaker enemy forces at some decisive point."[62]

Sun Tzu and Basil Liddell Hart, on the other hand, expounded the strategic thought of an "indirect approach." Such an approach, according to Sun Tzu, suggests that direct attacks on well-defended positions should be avoided at all costs.[63] Defeating an enemy, therefore, requires isolating and demoralizing them to ultimately break their will to resist rather than applying superior force.[64] Liddell Hart also developed his vision of an indirect approach after witnessing the horrors of the First World War (WWI). His conception did not exclude decisive battle, but instead, he proposed that "the aim of strategy must be to bring about this battle under the most advantageous circumstances."[65] Accordingly, he argued that a perfect strategy would be one wherein a decision would be produced without significant battle.[66] Liddell Hart, therefore, proposed his definition of strategy as "the art of distributing military means to fulfill the ends of policy, which takes into account the many alternative and complementary means other than battle which strategy possesses."[67] His view of strategy thus reflects grand strategy rather than military strategy.

Over time, further emphasis was placed on the notion that various means must be employed to achieve objectives. For example, John Fuller stated that:

> Even in the past, not a few battles have been won by surprise rather than by force of arms, and if battle can be won without suffering loss, surely this is the most economical, if not the most traditional, way of gaining the strategical object.[68]

Fuller thus believed that achieving strategic objectives should ideally take place without a significant physical battle but instead by employing ways and means that surprise an opponent to prevent them from achieving the strategic objective. T.E. Lawrence also envisioned achieving strategic goals as something that is preferably done without significant physical battle but rather "by pressing our advantages mathematical and psychological."[69] According to Liddell Hart, the very purpose of strategy is, therefore, "not so much to seek battle as to seek a strategic situation so advantageous that if it does not in itself produce the decision, its continuation by a battle is sure to achieve this."[70] In other words, military strategy is the art of creating a strategic position that leads to a competitive advantage for friendly forces. It will simultaneously influence an adversary to make it realize that accepting its disadvantage rather than employing force is in its best interest. Literature on strategic theory captures these concepts in a simple and elegant model of ends (i.e., the "what" by describing the achievement of desired political objectives), ways (i.e., the "how" in the form of a strategic approach), and means (i.e., the "with" that lists the resources to pursue the desired ends).[71]

Methodology

As described, exploiting the full potential of special operations depends on the skillful practice of strategy development and superior tactical behavior. This book is designed to determine how special operations fulfill their strategic promise. This will be done using an analytical framework consisting of three distinct parts appropriate to the strategic, operational, and tactical levels of war.

First, the research examines the validity of each case's strategy, assessing its suitability, feasibility, and acceptability:

- Suitability: could the operation be executed with the available resources?
- Feasibility: did achieving the objectives create the strategic effects that contributed to the desired end?
- Acceptability: did the strategic effects sought to justify the objectives pursued and consider tangible and intangible costs to achieve objectives?[72]

What follows is an assessment of the risks associated with the strategy. This assessment will include analyzing the consequences of success and failure; examining the ends, ways, and means relative to the strategic environment; and determining which strategic factors impacted the generation of strategic effects. The analytical framework's third and final part concerns tactical action. Detailed attention will be given to the TTPs employed against enemy vulnerabilities to gain temporal or sustained relative superiority. The results of the analysis will be particularly relevant when considering special operations from a multi-domain perspective and viewed from opposite sides of the board.

A first step in this direction is taking a closer look at the typology that characterizes most, if not all, special operations. They can be subdivided into two different types. The first type is special operations that follow a direct approach. Direct special operations are executed with surgical precision, often autonomously, and are typically short in duration. The character of these operations is typically offensive in nature, as they usually include a precision-strike capability. By conducting these operations, states aim to generate strategic effects, which, in turn, contribute directly to achieving their political objectives. Examples of direct special operations include the Israeli hostage rescue mission code-named Operation Thunderbolt in Entebbe (Uganda), the airstrike that killed Al-Qaeda leader Abu Musab al-Zarqawi in Iraq on June 7, 2006, and Operation Neptune Spear, which led to the death of Al-Qaida leader Osama Bin Laden on May 2, 2011. As described in Chapter 3 of this book, these endeavors involve a high degree of risk, and mission failure will be highly visible and dramatic. Chapters 3, 4, and 5 highlight that the planning and preparation phases of direct special operations may range from days to years. Chapters 3 and 4 illustrate that the execution phase is typically short in duration. Notably, with the possibility to conduct such operations through cyberspace, the notion of time has changed dramatically, as the domain allows for a protracted execution phase. See Chapter 5 for an example of this. Finally, target development and post-mission activities will require cooperation between various departments, agencies, and even countries, thereby emphasizing the need for a systems view of special operations. These aspects will be discussed in Chapters 4 and 5.

The second main type is indirect special operations that have a broad, geographically dispersed, and protracted character to exert influence in a broader politico-military campaign. This type of special operations is often referred to as special warfare, which includes military and paramilitary activities such as sabotage, subversion, and psychological operations requiring "to fight as guerrillas."[73] Examples of indirect special operations include the actions of T.E. Lawrence against the Ottoman Empire during WWI (1914–1918), the SOE in WWII (1940–1945), stay-behind operations in Western Europe during the Cold War (1947–1989), the British Special Air Service (SAS) in the Borneo Campaign (1963–1966), and US cooperation with the Afghan Northern Alliance to overthrow the Taliban regime in Afghanistan (2001). Indirect special operations fall between economic coercion and the overt use of military force, including direct special operations. Chapters 6 and 7 highlight that indirect special operations provide policymakers with an additional option to achieve political objectives while mitigating risk. These operations are typically employed in support of unconventional warfare (UW), counterinsurgency (COIN), and operational preparation of the environment (OPE). See Chapter 8 for an example. It is important to distinguish between indirect special operations and the previously described indirect approach to military strategy. While the distinctive element between direct and indirect special operations is the use of tactics, an indirect approach to military strategy differs from a direct approach through the selection of its target.

Research design

The starting point of any study should be defining a suitable research question.[74] Within this book, the research question is categorized as a "how" question: "how do the personnel involved fulfill the strategic promise of special operations, given their position within the larger framework of foreign policy?" According to Robert Yin, "how" questions can be answered using experiments, historical analysis, and case study research.[75] Additionally, a multiple case setup adds robustness to the findings, as Liddell Hart argues:

> The method in recent generations has been to select one or two campaigns, and to study them exhaustively as a means of professional training and as the foundation of military theory. But with such a limited basis the continual changes in military means from war to war carry the danger that our outlook will be narrow and the lessons fallacious.[76]

After combining these observations, a multiple case study design is the most suitable choice for this work because it offers the possibility to describe the differences and similarities between the selected special operations. Also, this design will present the reader with a broad outlook on the subject matter, and lessons are therefore grounded in both theory and practice.

Data collection and analysis

A book on special operations should ideally be written focusing on the direct environment of the missions themselves. However, too much focus on the operations would distract from a more abstract, in-depth investigation of special operations at large. Out of the vast number of special operations conducted at the time of this writing, six distinct case studies were selected based on three criteria. First, drawing from the definitions presented in this chapter, each mission was a "special operation" within its specific context. Second, following the typology of special operations presented above, the missions described were characterized as either direct or indirect special operations. The third selection criterion was accessibility: to develop a comprehensive understanding of the case studies, access to available data had to be guaranteed. Accordingly, following the emerging twenty-first century security environment, this book provides the reader with a look behind the scenes at some of its main competitors, including the Islamic Republic of Iran (Iran), PRC, Russian Federation (Russia), State of Israel (Israel), and the US.

Data collection took place during multiple field trips, including to Belgium, Germany, Israel, Japan, Lebanon, The Philippines, Ukraine, and the US. Interviewees included senior-level diplomats and military officers who were all directly involved with the operations. Other participants included journalists, academics, and private sector representatives. An essential aspect of the data is source protection, particularly given the risk of exposure. Therefore, all respondents remain anonymous, and the data is not traceable to a country, an individual, or a functional position.

The field trips offered unique insights into the planning, preparation, and execution of special operations, observed through the eyes of a distinct group of experts and practitioners. Therefore, this book contributes to the theory of special operations by enriching it with a strategic outlook on special operations in the twenty-first century that are planned, prepared, and executed by five key actors from across the globe.

For each case study, various operation orders, reports, and other relevant documentation were studied to create a more general insight into the mission. These reviews provided a more detailed overview of the organizational design of the mission, including its command-and-control structure. They enabled the selection of the personnel intimately involved with the mission for the interviews. Subsequently, an interview protocol was developed based on the insights obtained from document analysis.

Each case study is described in a separate chapter and is organized as follows: the first section provides a detailed overview of the background leading to the operation. Then, a brief description of the historical and geographical context of the area in which the operation took place is given. The sections that follow present a chronological description of the planning, preparation, and execution of the operation. To assess whether the operation fulfilled its strategic promise, the remainder of the chapter analyzes the alignment of policy, strategy, and tactics appropriate to the strategic, operational, and tactical levels of war. Each chapter concludes with a brief discussion of the main findings of the case study.

Reading guide

The remainder of this book consists of eight chapters. Chapter 2 covers the key characteristics of the twenty-first century strategic environment and how these affect special operations as a strategic instrument of foreign policy. This chapter concludes the theoretical part of the book and serves as the hinge where the focus shifts to its practical parts. Following the typology of special operations, part II comprises the first three case studies (Chapters 3, 4, and 5) representing direct special operations. Part III consists of three subsequent case studies (Chapters 6, 7, and 8) describing indirect special operations. Chapter 9 forms the final part of this book, summarizing its contributions to theory and practice.

Notes

1 Louis Allen, "Setting Europe Ablaze," *New Blackfriars* 48, no. 557 (1966): 35.
2 Richard W. Potter and Kalev I. Sepp, "The Renaissance Force: Selecting Soldiers and Forging Teams for Special Operations," in *Leadership: The Warrior's Art*, ed. Christopher Kolenda (Carlisle: Army War College Foundation Press, 2001).
3 Joint Special Operations University, *Special Operations Forces Reference Manual*, 5th ed. (MacDill AFB, Florida: JSOU Press, November 2023), 2.
4 Colin S. Gray, *Explorations in Strategy* (Westport, CT: Praeger, 1998).
5 Gray, 168.
6 Gray, 174.
7 Gray, *Explorations in Strategy*; Bernd Horn, "The Strategic Utility of Special Operations Forces," *Canadian Military Journal* 14, no. 4 (2014): 66–70.
8 Joseph Lutz, "Discussants," in *Special Operations in US Strategy*, ed. Frank Barnett, Hugh Tovar, and Richard Schultz (Washington, D.C.: National Defense University Press, 1984), 47.
9 Alfred H. Paddock, Jr., *US Army Special Warfare: Its Origins* (Washington, DC: National Defense University Press, 1982), 31.
10 Zhang Yuliang, *ed. The Science of Military Campaigns* (Beijing: National Defense University Press, 2006), 219.
11 Yuliang, *The Science of Military Campaigns.*
12 Ibid.
13 Ibid.
14 Sun Tzu, *The Art of War*, trans. Samuel B. Griffith (New York: Oxford University Press, 1963), 229.
15 John J. Dziak, "The Soviet Approach to Special Operations," in *Special Operations in US Strategy*, ed. Frank R. Barnett, B. Hugh Tovar, and Richard H. Shultz (New York: National Defense University Press, 1984).
16 Dziak, "The Soviet Approach."
17 Ibid.
18 Christopher Marsh, *Developments in Russian Special Operations: Russia's Spetsnaz, SOF and Special Operations Forces Command* (Ottawa: CANSOFCOM Education & Research Centre, 2017).
19 Lester Grau and Charles Bartles, *The Russian Way of War: Force Structure, Tactics, and Modernization of the Russian Ground Forces* (Fort Leavenworth, KS: Foreign Military Studies Office, 2016).
20 Alexey Nikolsky, "Little, Green and Polite: The Creation of Russian Special Operations Forces," in *Brothers Armed: Military Aspects of the Crisis in Ukraine*, ed. Colby Howard and Ruslan Pukhov (Minneapolis: East View Press, 2015).
21 Nikolsky, "Little, Green and Polite."

22 Ministerstvo Oborony Rossiiskoi Federatsii, Formy primeneniya Vooruzhyonnykh cil Rossiiskoi Federatsii (Moscow: Ministerstvo Oborony Rossiiskoi Federatsii, n.d.).

23 Alexander Sladkov, "Novyi Rossiiskii Spetsnaz dlya zagranitsy," Television Channel "Rossiya 24", program "Vesti Nedeli," (April 28, 2013).

24 Aleksey Nikolsky, "Russian 'Spetsnaz' Forces – From Saboteurs to Court Bailiffs," *Moscow Defense Brief*, 40, no. 2 (2014): 126–127.

25 Maurice Tugwell and David Charters, "Special Operations and the Threats to United States Interests in the 1980s," in *Special Operations in US Strategy*, ed. Frank Barnett, Hugh Tovar, and Richard Schultz (Washington, D.C.: National Defense University Press, 1984), 5.

26 Gray, *Explorations in Strategy*.

27 NATO, glossary of terms and definitions AAP-06 (NATO Standardization Agency, 2021), 97.

28 NATO, 35.

29 Ibid., 26.

30 NATO, Allied Joint Doctrine for Special Operations AJP-3.5(B).

31 DOD, *Joint Publication 3-05 Special Operations* (Washington, DC: Department of Defense, 2014), ix.

32 James D. Kiras, *Special Operations and Strategy: From World War II to the War on Terrorism* (New York: Routledge, 2006), 5.

33 John Arquilla, *ed. From Troy to Entebbe: Special Operations in Ancient and Modern Times* (Lanham, Maryland: University Press of America, 1996).

34 Carl von Clausewitz, *On War*, ed. and trans. Michael Howard, Peter Paret, and Bernard Brodie (Princeton, N.J.: Princeton University Press, 1984), 119.

35 William H. McRaven, *Spec Ops Case Studies in Special Operations Warfare* (New York: The Random House Publishing Group, 1995), 1.

36 McRaven, *Spec Ops*.

37 Robert G. Spulak Jr., *A Theory of Special Operations: The Origin, Qualities, and Use of SOF* (Hurlburt Field, FL: Joint Special Operations University Press Publications, 2007).

38 Spulak, 25.

39 Spulak, *A Theory of Special Operations*.

40 Lieutenant General (Ret.) Eric Wendt, "Comprehensive Defense: A Whole-of-Society Approach via Irregular Forces," *Special Warfare* 34, no. 2 (2021).

41 Wendt, 4.

42 Wendt, 4.

43 Lieutenant General (Ret.) Eric Wendt, interview by author, Monterey, November 21, 2022.

44 Wendt, interview.

45 Steve Lambakis, "Colin Gray on the strategic utility of special operations," *Comparative Strategy* 40, no. 2 (2021): 208; Spulak, *A Theory of Special Operations*.

46 Colin S. Gray, "Handfuls of Heroes on Desperate Ventures: When do Special Operations Succeed?" *Parameters* 29, no. 1 (1999): 2–24.

47 Gray, "Handfuls of Heroes."

48 Arquilla, xv–vxi.

49 Lambakis, "Colin Gray"; Horn, "The Strategic Utility."

50 Henry A. Murray et al. *Assessment of Men: Selection of Personnel for the Office of Strategic Services* (New York: Rinehart and Co., 1948), 11.

51 Gray, *Explorations in Strategy*.

52 Samuel P. Huntington, *The Soldier and the State: The Theory and Politics of Civil-Military Relations* (Boston: Belknap of Harvard University Press, 1957).

53 Colin S. Gray and Jeannie L. Johnson, "The Practice of Strategy," in *Strategy in the Contemporary World: An Introduction to Strategic Studies*, 6th ed., ed. John Baylis, James Wirtz, Eliot Cohen, Colin S. Gray (New York: Oxford University Press, 2019).

54 Gray, *Explorations in Strategy.*
55 Colin S. Gray, "Strategic Sense – Missing from Action," *Infinity Journal* 5, no. 3 (2016): 4–8.
56 Gray and Johnson, 392.
57 Ibid., 392.
58 Clausewitz, *On War*, 258.
59 Ibid, 194–197.
60 Ibid, 194–197.
61 Ibid, 194–197.
62 John Shy, "Jomini," in *Makers of Modern Strategy: From Machiavelli to the Nuclear Age*, ed. Peter Paret (Princeton, NJ: Princeton University Press, 1986), 146.
63 Sun Tzu, *The Art of War.*
64 Ibid.
65 Basil Liddell Hart, *Strategy*, 2nd ed. (New York: Penguin Books, 1991), 324.
66 Liddell Hart, *Strategy.*
67 Ibid, 324.
68 John F.C. Fuller, *On Future Warfare* (London: Sifton, Praed & Co., Ltd, 1928).
69 Thomas E. Lawrence, *Revolt in the Desert* (New York: Doran, 1927), 66.
70 Liddell Hart, 338–339.
71 Arthur F. Lykke Jr., "Defining Military Strategy," *Military Review* 69, no. 5 (1989); Harry R. Yarger, *Strategic Theory for the 21st Century: The Little Book on Big Strategy* (Carlisle, PA: US Army War College, Strategic Studies Institute, February 2006).
72 Yarger, *Strategic Theory for the 21st Century.*
73 Paddock, Jr., *US Army Special Warfare.*
74 Robert K. Yin, *Case Study Research Design and Methods*, 5th ed. (Thousand Oaks: Sage Publications, 2014), 1.
75 Yin, *Case Study Research.*
76 Liddell Hart, 4.

Bibliography

Allen, Louis. "Setting Europe Ablaze." *New Blackfriars* 48, no. 557 (1966): 35–38.

Arquilla, John ed., *From Troy to Entebbe: Special Operations in Ancient and Modern Times* (Lanham, Maryland: University Press of America, 1996).

Clausewitz von, Carl, *On War*, ed. and trans. by Michael Howard, Peter Paret, and Bernard Brodie (Princeton, N.J.: Princeton University Press, 1984).

Department of Defense, *Joint Publication 3-05 Special Operations* (Washington, DC: Department of Defense, 2014), https://irp.fas.org/doddir/dod/jp3_05.pdf.

Dziak, John J., "The Soviet Approach to Special Operations," in *Special Operations in US Strategy*, ed. Frank R. Barnett, B. Hugh Tovar, and Richard H. Shultz (New York: National Defense University Press, 1984).

Fuller, John F.C. *On Future Warfare* (London: Sifton, Praed & Co., Ltd, 1928).

Grau, Lester, and Bartles Charles. *The Russian Way of War: Force Structure, Tactics, and Modernization of the Russian Ground Forces* (Fort Leavenworth, KS: Foreign Military Studies Office, 2016).

Gray, Colin S. *Explorations in Strategy* (Westport: CT, Praeger, 1998).

Gray, Colin S. "Handfuls of Heroes on Desperate Ventures: When Do Special Operations Succeed?," *Parameters* 29, no. 1 (1999): 2–24.

Gray, S. Colin, and L. Johnson Jeannie, "The Practice of Strategy," in *Strategy in the Contemporary World: An Introduction to Strategic Studies*, 6th ed., ed. John Baylis, James Wirtz, Eliot Cohen, Colin S. Gray (New York: Oxford University Press, 2019).

Horn, Bernd. "The Strategic Utility of Special Operations Forces,"*Canadian Military Journal* 14, no. 4 (2014): 66–70.

Huntington, Samuel P. *The Soldier and the State: The Theory and Politics of Civil-Military Relations* (Boston: Belknap of Harvard University Press, 1957).
Joint Special Operations University, *Special Operations Forces Reference Manual*, 5th ed. (MacDill AFB, Florida: JSOU Press, November 2023), https://info.publicintelligence.net/JSOU-SOF-Reference.pdf.
Kiras, James D. *Special Operations and Strategy: From World War II to the War on Terrorism* (New York: Routledge, 2006).
Lambakis, Steve. "Colin Gray on the Strategic Utility of Special Operations," *Comparative Strategy* 40, no. 2 (2021): 205–208.
Lawrence, Thomas E. *Revolt in the Desert* (New York: Doran, 1927).
Liddell Hart, Basil., *Strategy*. 2nd ed. (New York: Penguin Books, 1991).
Liddell Hart, Basil. *Strategy of the Indirect Approach* (London: Faber and Faber Limited, 1941).
Lutz, Joseph, "Discussants," in *Special Operations in US Strategy*, ed. Frank Barnett, Hugh Tovar, and Richard Schultz (Washington, D.C.: National Defense University Press, 1984).
Lykke, Arthur F Jr., "Defining Military Strategy," *Military Review* 69, 5 (1989).
Marsh, Christopher. *Developments in Russian Special Operations: Russia's Spetsnaz, SOF and Special Operations Forces Command* (Ottawa: CANSOFCOM Education & Research Centre, 2017), https://publications.gc.ca/site/eng/9.831086/publication.html.
McRaven, William H. *Spec Ops Case Studies in Special Operations Warfare* (New York: The Random House Publishing Group, 1995).
Ministerstvo Oborony Rossiiskoi Federatsii, Formy primeneniya Vooruzhyonnykh cil Rossiiskoi Federatsii (Moscow: Ministerstvo Oborony Rossiiskoi Federatsii, n.d.), http://encyclopedia.mil.ru/encyclopedia/dictionary/details.htm?id=14014.
Murray, Henry A., Donald W. MacKinnon, James G. Miller, Donald W. Fiske, and Eugenia Hanfmann. *Assessment of Men: Selection of Personnel for the Office of Strategic Services* (New York: Rinehart and Co, 1948).
NATO, glossary of terms and definitions AAP-06. NATO Standardization Agency, 2021.
NATO, Allied Joint Doctrine for Special Operations AJP-3.5(B).
Nikolsky, Alexey, "Little, Green and Polite: The Creation of Russian Special Operations Forces," in *Brothers Armed: Military Aspects of the Crisis in Ukraine*, ed. Colby Howard and Ruslan Pukhov (Minneapolis: East View Press, 2015).
Nikolsky, Aleksey, "Russian 'Spetsnaz' Forces – From Saboteurs to Court Bailiffs," *Moscow Defense Brief* 40, 2 (2014), http://mdb.cast.ru/mdb/1-2014/item4/article2/. (Not accessible as of [30/07/2024])
Paddock, Alfred H. Jr. *US Army Special Warfare: Its Origins* (Washington, DC: National Defense University Press, 1982).
Potter, Richard W., and I. Sepp Kalev, "The Renaissance Force: Selecting Soldiers and Forging Teams for Special Operations," in *Leadership: The Warrior's Art*, ed. Christopher Kolenda (Carlisle: Army War College Foundation Press, 2001).
Searle, Tom. "Outside the Box: A New General Theory of Special Operations." *JSOU Report* 17, 6 (2017): 103–117, https://jsou.edu/Press/PublicationDashboard/59.
Shy, John, "Jomini," in *Makers of Modern Strategy: from Machiavelli to the Nuclear Age*, ed. Peter Paret (Princeton, NJ: Princeton University Press,1986).
Sladkov, Alexander, "Novyi Rossiiskii Spetsnaz dlya zagranitsy," Television Channel "Rossiya 24", program "Vesti Nedeli," (April 28, 2013).
Spulak, Robert Jr., "*A Theory of special operations: The Origin, Qualities, and Use of special operations forces*," JSOU Report 07-7, October 2007: 1–63, https://apps.dtic.mil/sti/pdfs/ADA495521.pdf.
Tugwell, Maurice, and Charters David, "Special Operations and the Threats to United States Interests in the 1980s," in *Special Operations in US Strategy*, ed. Frank Barnett, Hugh Tovar, and Richard Schultz (Washington, D.C.: National Defense University Press, 1984).
Tzu, Sun, *The Art of War*, trans. by Samuel B. Griffith (New York: Oxford University Press, 1963).

Watling, Jack, "Sharpening the Dagger: Optimising Special Forces for Future Conflict," RUSI Whitehall Report, 2021.
Wendt, Eric. "Comprehensive Defense: A Whole-of-Society Approach via Irregular Forces," *Special Warfare* 34, no. 2 (2021): 4.
Yarger, Harry R. *Strategic Theory for the 21st Century: The Little Book on Big Strategy* (Carlisle, PA: US Army War College, Strategic Studies Institute, February 2006).
Yin, Robert. K., *Case Study Research Design and Methods*, 5th ed. (Thousand Oaks: Sage Publications, 2014).
Yuliang, Zhang, ed., *Tshe Science of Military Campaigns* (Beijing: National Defense University Press, 2006).

2 The Strategic Environment

Introduction

Significant political, economic, social, and technological changes characterize the twenty-first century strategic environment. These changes present challenges and opportunities for special operations. For example, various state and non-state actors are trying to expand their political, military, economic, and societal power and influence. Three superpowers, namely the People's Republic of China (China), the Russian Federation (Russia), and the United States (US), play key roles in this new era of strategic competition. From a military perspective, special operations are being conducted across all domains (i.e., air, land, sea, space, and cyberspace), generating effects in three dimensions (i.e., physical, virtual, and cognitive). Covering all domains, geographical space will become compressed, while the operating speed will significantly increase. Furthermore, the operational environment (OE) will be contested, as a myriad of sensors will lead to ubiquitous surveillance. These dynamics are changing the conduct of war, which has significant implications for the planning, preparation, and execution of special operations.

The changing conduct of war

Warfare is as old as humanity itself. However, its conduct has continuously changed since the first war, believed to have occurred between 4000 and 3500 BCE (Before the Common Era) in ancient Mesopotamia. Indeed, Martin van Creveld argues that:

> Each time advancing technology enabled mankind to move into a new environment – from the land into the surface of the sea, from the surface of the sea into the depths on one hand and the air and outer space on the other – war quickly followed.[1]

The battle of Marathon (490 BCE), the integration of sail and gun during the first three Anglo–Dutch Wars (1652–1674), the birth of electronic intercepts in the time of the Boer War (1899–1902), the development of air warfare throughout the twentieth century, and the use of malicious code in today's shadowy cyberwars are just some contextual illustrations of how the conduct of war has changed over time.

DOI: 10.4324/9781003365112-3

Every generation is thus confronted with its own type of warfare dictated by its specific historical context. Carl von Clausewitz states that each type of warfare has "its own limiting conditions, and its own peculiar preconceptions."[2] The next generations will create their own conduct of war because of responses to significant political and technological changes in waging war.

Today, a global innovation model drives the proliferation of emerging technologies. Examples of emerging technologies are robotic and autonomous systems (RAS), unmanned aerial systems (UAS), hypersonic systems, and information capabilities connected to cyberspace combined with quantum computing, artificial intelligence (AI), machine learning (ML), and nanotechnology. These technological developments have merged the physical, virtual, and cognitive dimensions of the OE and continue to compress the operational factors of time, space, and force. Moreover, these technologies, according to Henry Kissinger, are developing faster than military thinkers can fully understand their impact on modern warfare:

> The pervasiveness of networked communications in the social, financial, industrial, and military sectors has vast beneficial aspects; it has also revolutionized vulnerabilities.... Internet technology has outstripped strategy or doctrine – at least for the time being. In the new era, capabilities exist for which there is as yet no common interpretation – or even understanding.[3]

What we do know, however, is that these technologies might create so-called cascading effects where information about an event in one part of the world is being shared in real-time with those far away. For example, in 2008, Sohaib Athar, an ordinary Pakistani citizen, was in Abbottabad, where he unwittingly reported live on Operation Neptune Spear via his Twitter account.[4] On hearing helicopters, Athar tweeted, "Helicopter hovering above Abbottabad at 01:00 (is a rare event)."[5] Shortly afterward, Athar tweeted, "A huge window-shaking bang here in Abbottabad. I hope it is not the start of something nasty."[6] This example clearly illustrates that networked communications combined with the open innovation model have increased the speed and enhanced the geographic reach of information.

The technical application of information could generate significant effects (including collateral effects) in the three dimensions of the OE (i.e., physical, virtual, and cognitive). As these technological developments mature and military applications become more apparent, their impact will present significant challenges and opportunities for special operations. These challenges and opportunities include enhancing reach, gaining access, maintaining freedom of movement, and influencing the OE. Technological developments, however, do not achieve a competitive advantage in isolation but require conceptual and organizational innovation to enable their operationalization effectively.[7]

Given these changes in the strategic environment, special operations will require a different notion of their performance. This notion should answer the overarching question of how they could remain a key strategic tool while moving forward and encountering the complex twenty-first century security challenges. Indeed, the changing characteristics of warfare will require that decisiveness in special operations will increasingly be determined by the ability to achieve cross-dimensional effects on the military objectives present in the OE and, if required, assuring that the operation remains undetected and its sponsor unattributed. To avoid detection and attribution, special operations must be conducted covertly or clandestinely in multiple domains of military operations, sequentially or simultaneously.[8] Over the last century, military operations across multiple domains have developed in theory and practice.[9] German *Blitzkrieg*, US AirLand Battle, China's Informatized Wars, Russian *Gibridnaya Voyna*, and US Multi-Domain Operations (MDO) are examples that illustrate how military strategies and combined arms warfare have emerged once new domains of military operations became accessible.[10]

More than three decades ago, John Arquilla and David Ronfeldt predicted that "Cyberwar" was coming and would have a significant impact on the twenty-first century OE, similar to the implications of combined arms warfare during the twentieth century.[11] Their predictions could not have been more accurate, as today, the integration of the cyberspace domain, with its emphasis on the Internet and computer networks, has become a critical component of warfare.[12] As a result, the planning, preparation, and execution of special operations are increasingly subject to adversarial threats emerging from information warfare, requiring defensive capabilities while simultaneously necessitating the use of information in an offensive manner.[13] From a Western perspective, the integration of cyberspace within the other four domains of military operations – air, land, sea, and space plus the electromagnetic spectrum (EMS) – is still at a developmental stage.[14] In fact, how cyberspace capabilities are integrated into the joint functions of most Western military organizations faces significant constraints because of the authorities, roles, and responsibilities of the various agencies involved.[15] For instance, accessing a target within cyberspace is affected by legal, political, or operational limitations. Consequently, gaining virtual access to cyberspace is not merely a matter of capability but rather subject to high-level political decision-making and legislation. Additionally, generating effects in cyberspace can be hindered by specific characteristics of the cyberspace domain itself used by a myriad of friendly, neutral, and hostile actors simultaneously. This multi-actor composition requires detailed planning and close coordination on deconflicting and synchronizing actions with non-military actors.[16] The utilization of the cyberspace domain does not fit the way in which special operations are supposed to be planned, prepared, organized, and executed to meet the challenges and exploit the opportunities globally presented by the strategic environment.

Within the changing conduct of war, special operations have reached a new inflection point, as was emphasized by Rear Admiral (Ret.) Hugh Howard. [17] Howard refers to this inflection point as entering "the fifth modern era of special operations."[18] According to Lieutenant General James Slife, such a transformation, seen as a new inflection point, will enable US Special Operations Forces (SOF) to remain relevant in the emerging OE.[19] Special operations' first inflection point refers to Operation Eagle Claw (1979), a complex military effort to rescue American hostages in Tehran that failed.[20] As a result, both US Special Operations Command (USSOCOM) and Joint Special Operations Command (JSOC) were established within the US Department of Defense (US DOD).[21] The second inflection point presented itself shortly after the devastating 9/11 terrorist attacks when US SOF became the primary tool in the global war on terrorism. With the emergence of strategic competition, special operations have arrived at a third inflection point.[22] For special operations, such inflection points imply a moment of significant change to remain a key strategic tool while moving forward and encountering the complex security challenges of their time.

Great power rivalry

As we move forward in the twenty-first century, strategic competition presents an era wherein various state and non-state actors seek to advance their interests. The presence of three major centers of power (i.e., China, Russia, and the US) is known as multipolarity and differs from the bipolar world (i.e., two centers of power), the period between 1947 and 1991, known as the Cold War. Robert Kaplan has identified various similarities between the Cold War and the current multipolar era of strategic competition, including global power projection, military arms races, technological competition, and ideological aspects. Therefore, Kaplan defines this new era as the "new Cold War."[23] Whether Kaplan is correct or not, the new multipolar worldview characterized by hegemonic rivalry has significant implications for special operations, as the various directions into which this rivalry might develop will lead to cooperation, competition, or conflict between the three powers.

The possible interrelations between the three great powers, seen from a more global perspective, can be described through three different theoretical lenses, namely realism, liberalism, and constructivism. Realism offers a first perspective focusing on power and influence. A fundamental assumption of hegemonic war and international change is that the dominant hegemon and rising state compete to redistribute power.[24] Furthermore, history has proven that rising hegemons have often been seen as troubleshooters challenging territorial boundaries and often feel that things were unjustifiably taken from them.[25]

Contrary to this realist perspective, liberalism's view on the role of economic interdependence as a pacifying power explains how the dominant and challenging hegemons might avoid conflict. A fundamental assumption of liberalism is the belief that bilateral economic trade leads to mutually beneficial relationships between states.[26] Moreover, the more economic trade flows between nations, the greater the likelihood that all sides will be keen on maintaining peaceful relations.[27]

Constructivism, on the other hand, assumes that social constructs that a group of people and its leaders take for granted shape international relations.[28] Social constructs are considered subjective and include identities, strategic cultures, and norms that are not universal but rather prevalent within a specific group or nation.[29] These social constructs reflect the perceptions of the dominant and rising hegemons about each other.[30]

The three international relations theories provide distinct possible outcomes of strategic competition. From a military perspective, however, all three great powers clearly emphasize their intent to isolate flashpoints. For example, one core tenet of US MDO is calibrated force posture and penetrating layered stand-off to gain operational access and to contest local conflict areas.[31] The People's Liberation Army (PLA) refers to "local wars under informatized conditions."[32] Finally, Russia is continuously challenging the boundaries of escalation management by seeking to expand its operational activities before potentially passing the threshold of nuclear conflict.[33] If the intentions to isolate flashpoints will manifest, a global confrontation between the three great powers is unlikely. Strategic competition, therefore, may result in irregular warfare, using all instruments of national power to gain a competitive advantage over supply chains, natural resources, and extraterritorial interests to provide political and economic leverage. An example of such a potential irregular conflict could be found in countries rich in natural resources such as cobalt, an essential ingredient in lithium battery production. The largest cobalt reserves in the world (+/− 70 percent) are in the Democratic Republic of the Congo (DRC), a fragile nation in Central Africa ravaged by decades of armed conflict. While there may be enough cobalt in the DRC mines for the US, there certainly is not enough for the other two superpowers. The American writer Mark Twain once said that history never repeats itself, but it often does rhyme. If he is correct, the energy transition could be much more violent than what it seeks to replace. A comparison with the discovery of oil by George Reynolds in Persia (modern-day Iran) in 1908 is easy to make and recalls the many conflicts of the twentieth century fueled by the "Black Gold." On the other hand, when flashpoints escalate into conflict on a broader scale, conventional or even nuclear war between the great powers is not unthinkable.

In these two distinct types of conflict (i.e., conventional and irregular), special operations are a strategic instrument of statecraft to change the balance of military forces. Examples of special operations to shift the balance of forces include Operation Source (September 1943), the Allied attempt to sink the German Battleship "*Tirpitz*" while anchored in *Kåfjord*, northern Norway, and the Italian Raid on Alexandria harbor in Egypt (December 1941) to disable the Royal Navy battleships "Queen Elizabeth" and "Valiant."[34] A second significant purpose of special operations is often to achieve a solution in the near term to allow for developing a long-term solution. In other words, by employing special operations, policymakers "buy time." Examples of special operations that were conducted "to by time" include Operation *Tungtvannsaksjonen* (1940–1944), during which Norwegian commandos sabotaged German heavy water production via hydroelectric plants, and the assassinations of Iranian nuclear scientists (2010–2012), unofficially attributed to Israel. Special operations' final strategic purpose is to significantly expand the reach over which military force can be employed. Their small and light footprint makes them an ideal instrument to be deployed (quickly) over great distances without robust logistical support. Examples of such special operations include the initial phase of Operation Enduring Freedom (2001–2014), during which US SOF and operations officers from the Central Intelligence Agency (CIA) worked closely with the Afghan Northern Alliance to overthrow the Taliban regime. Also, the numerous strikes that US SOF conducted deep in Syrian territory to kill or capture the Islamic State of Iraq and Syria (ISIS) senior leadership (2014–today) are clear examples of expanding the reach over which military force can be employed.

From a military perspective, conventional and irregular warfare are increasingly taking place across all domains of military operations. Access to, and freedom of movement in these domains are critical conditions to generate cross-dimensional effects in the OE. This is traditionally known as combined arms warfare, or joint operations, in military literature.[35] A more recent Western concept is MDO, which can be viewed in broad (competition) and narrow (conflict) terms. According to the North Atlantic Treaty Organization (NATO), the broad concept of MDO encompasses the employment of all instruments of national power to achieve governmental and non-governmental objectives.[36] The narrow view of MDO, also called Joint All Domain Operations (JADO), is defined in US military doctrine as operations occurring in the air, land, maritime, cyberspace, and space domains, plus the EMS.[37] The narrow view of MDO will be applied in the remainder of this book.

One could argue whether MDO is a new concept of warfare or merely an expansion of combined arms warfare, as both concepts share several significant similarities. For example, the US Army defined combined arms warfare as "the basic idea that different arms and weapons systems must be used in concert to maximize the survival and combat effectiveness of each other."[38] The doctrine lists several combined arms derived from the air, land, and sea domains and the EMS.[39] The central idea behind MDO is quite similar to that of combined arms warfare, as the concept seeks to integrate military operations conducted in the traditional domains of air, land, and sea, and in the twenty-first century, obviously also in cyberspace, space, and the EMS.[40] Accordingly, MDO has expanded the possibilities of combined arms warfare as new domains became available and applicable to military operations.

The operational environment

Special operations are planned, prepared, and executed in the five distinct domains of military operations. They are listed in military doctrine and operating concepts: air, land, sea, space, and cyberspace.[41] The domains are defined in US doctrine as "a sphere of activity or influence with common and distinct characteristics in which a force can conduct joint functions."[42] The domains are distinct in nature yet interrelated, and this is reflected in their hierarchical geophysical nature as outlined by General (Ret.) Larry Welch:

> The hierarchy begins with the land surface of the earth surrounded by the maritime domain. All the land and seas are surrounded by the air domain, and the air domain is surrounded by the space domain. In contrast to the other domains, cyberspace is embedded in all domains and operation in all domains is dependent on operation in cyberspace. Hence, military operations in all domains depend on operations in, through, and from cyberspace.[43]

Just as Welch perfectly explained the dependence of military operations on cyberspace, the opposite is also true, as maneuvers in or through cyberspace rely on infrastructure located in the physical domains.[44] Furthermore, US doctrine describes that "the relationship between space and cyberspace is unique in that virtually all space operations depend on cyberspace, and a critical portion of cyberspace bandwidth can only be provided via space."[45] These interrelations between the domains of military operations indicate that special operations increasingly depend on having access to and freedom of maneuver across domains to gain a competitive advantage over an opponent.

Contrary to the three physical domains of air, land, and sea, cyberspace is not confined by geographic and political boundaries. The domain also plays an important role in the control of critical infrastructure sectors, including water systems, health care, transportation, and energy. Moreover, cyberspace is human-built, dominated by the private sector, and hence subject to open innovation. As a result, the domain is influenced by rapid and continuous change and development.

The EMS is not part of the official domains of military operations but is considered an element of the OE.[46] US military doctrine states the EMS as "a maneuver space essential for facilitating control within the operational environment (OE) and impacts on all portions of the OE and military operations."[47] To further complicate matters, the doctrine refers to the Information Environment (IE) that:

> Comprises and aggregates numerous social, cultural, cognitive, technical, and physical attributes that act upon and impact knowledge, understanding, beliefs, world views, and, ultimately, actions of an individual, group, system, community, or organization. The information environment also includes technical systems and their use of data. The information environment directly affects all OEs [operational environments].[48]

In this completely new IE, the concepts of misinformation (false or misleading information) and disinformation (deliberate false information purposely spread to deceive) deserve special consideration, as they may not only disrupt the regular collection of information but also may interfere with the correct knowledge in society at large. When it comes to the execution of special operations in multiple domains of military operation, the activities automatically take place in the IE, as it is affected by all OEs. Coping with misinformation and/or disinformation has thus become an important aspect in the planning and execution of special operations. Figure 2.1 illustrates the interrelatedness between the five domains and the EMS.

Next to the domains of military operations, the OE consists of three distinct dimensions, namely physical, virtual, and cognitive. The physical dimension is where human life and physical infrastructure exist. It is typically associated with the employment of physical capabilities such as weapon systems and their destructive effects. The virtual dimension conveys data and information between people through Information Systems (IS), networks, and global interconnectivity. It is connected to the employment of virtual capabilities, including computer code. The cognitive dimension refers to a person, a group, or an organization and their norms, values, beliefs, culture, interests, objectives, and interactions between them. It is related to employing cognitive capabilities such as message content (i.e., message payload) and its interaction with human consciousness, including thought,

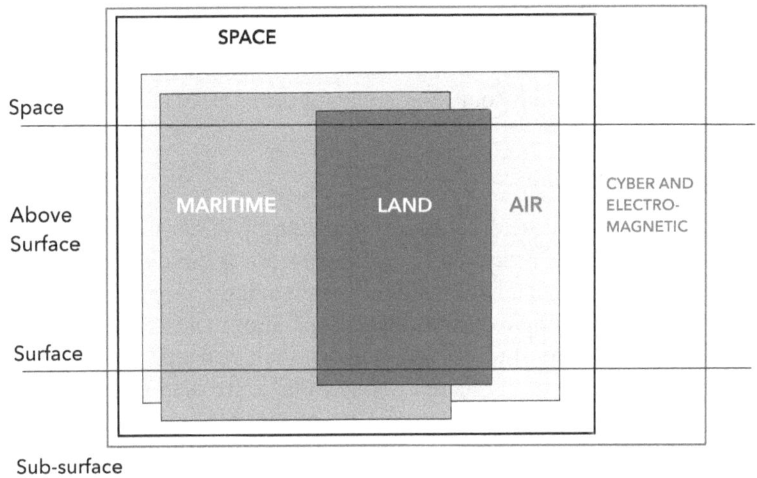

Figure 2.1 Relationships between the domains of military operations and EMS.

perception, emotion, and will. To generate effects from cross-domain maneuvers, special operations need to focus on the interplay between these three dimensions. In Western special operations concepts and doctrines, the ability to generate effects in the cognitive dimension is becoming increasingly important, as was stated by Lieutenant General Jonathan Braga:

> Information advantage and information operations – I think we are watching it daily, the strategic impact this has. I cannot envision a future where that does not increase in importance, affecting targeting audiences, general populations, governments, armies, morale, and eroding their overall effectiveness.[49]

For the planning, preparation, and execution of special operations, information is becoming increasingly important, and thus, cyberspace. Indeed, cyberspace enables the employment of virtual capabilities that could generate significant effects in all three dimensions of the OE, particularly in the cognitive dimension. Furthermore, the growing importance of cyberspace as part of kinetic operations was emphasized by Howard, saying "with cyber and electronic warfare, with our proximity to access to hard targets, we see ourselves as a part of that kill chain, in extending the reach of the cyber and electronic warfare enterprises."[50] Accordingly, cyberspace should become an integral element of special operations. Indeed, the PLA already included "special technical operations" in its military doctrine and emphasized that computer network operations, with their potential for more excellent concealment, have become a critical component of Chinese special operations.[51]

This approach changes our understanding of the OE confined by geographical boundaries. This newly created operational area is characterized by an expansion that transcends our long-term understanding of a traditional area of operations (AO) composed of the rear, close, and deep areas. Indeed, special operations may include the synergistic employment of capabilities across all domains, at the three levels of warfare, and throughout geographic locations across the globe.

Detection and attribution

The updated OE is contested across all domains of military operations due to emerging technologies. Consequently, gaining access to these domains and maintaining freedom of action has become highly challenging. Competitors increasingly disrupt activities, including exposing special operations and the people executing these missions. Exposure of forces will degrade their potential to deploy covertly or clandestinely in the future, thereby effectively decreasing their strategic utility. Exposing a competitor's operators is a recent trend in strategic competition. It finds its roots in cyberspace, where various Western countries explicitly indicted and sanctioned Chinese and Russian forces responsible for cyber espionage and subversion.[52] Throughout the years, exposing operatives has expanded to personnel conducting special operations in the traditional air, land, and sea domains, including operatives of Iran's Islamic Revolutionary Guard Corps (IRGC) Quds Force.[53] Western nations also started to expose (primarily) Russian operations and their operatives, as observed in the hack of Advanced Persistence Threat (APT) 29 "Cozy Bear."[54] Finally, as in Libya, the US has also started reporting Russian (proxy) activities in third countries.[55]

The increased risk of exposure requires a different interpretation of detection and attribution thresholds relative to the density of the contested OE. The density level is non-linear and depends on an actor's Intelligence, Surveillance, and Reconnaissance (ISR) capabilities and Anti-Access Aerial Denial (A2/AD) systems across domains. At the same time, Russian and US operational concepts describe special operations as the primary ways to infiltrate these contested environments, find and fix the critical components of these systems, and finish them by providing terminal guidance or traditional sabotage.[56] This task, however, seems like a paradox since SOF heavily rely on their clandestine posture while conducting Special Reconnaissance (SR) and thus must avoid detection at all costs. This is particularly relevant in today's contested OE, where detection often leads to attribution, as we have learned from various special operations attributed to their sponsor after the operation was detected. Examples of this include the Israeli operation in the United Arab Emirates (UAE), during which the Mossad killed senior Hamas official Mahmoud Al-Mabhouh on January 19, 2010, and the Russian operation in the United Kingdom (UK) in which Sergej Skripal, a former operative of Russia's Main Intelligence Directorate (GRU) and his daughter Yulia were poisoned on March 4, 2018, known as the Salisbury case.[57]

Therefore, the first challenge to conduct covert or clandestine special operations relates to the maneuvers in the contested physical domains. Traditional insertion and infiltration methods associated with the air and land domains, such as High-Altitude High Opening/High-Altitude Low Opening (HAHO/HALO) parachute infiltrations, as well as Rotary Wing (RW) assets insertions at low altitudes, are increasingly vulnerable to detection caused by the proliferation of airborne active electronically scanned array radar.[58] The same limitations apply to long-range vehicle infiltrations that are most effective in remote open areas while avoiding more densely populated areas, limiting their maneuverability.[59] Ground Moving Target Indication (GMTI) radar can cover large, open-space areas, making these ground movements highly vulnerable to detection.[60] Insertion by submarine is less susceptible to detection but is limited to coastal regions and poses constraints in the number of forces and equipment that can be inserted. However, developments in acoustic sonar systems indicate that sub-surface detection capabilities are improving.[61] Finally, while operating in "plain sight" with a cover identity, applying AI in combination with various biometric capabilities, such as face recognition, will significantly affect attempts to remain undetected and avoid attribution.[62]

The proliferation of Electronic Warfare (EW), whereby automated stand-off detection systems are increasingly contesting the EMS, forms a significant challenge to the command and control of special operations. Despite emerging technologies, detecting and pinpointing short communications transmitted over an unusual frequency remain challenging.[63] To elaborate as an example, more sensitive and precise monitoring is possible when detecting a particular transmission. The interceptor may also jam detected frequencies to impede transmissions.[64] However, a more effective tactic from the interceptor's position would be to conduct limited jamming, forcing the transmitter to boost its power level, leading to an increased electronic signature and, therefore, a higher probability of detection. Alternatively, the transmitting party may shift its transmissions to another frequency band of the EMS. This mode of operation is known as narrowing the frequency bands and forces the interceptor to narrow its focus for monitoring.

Once a transmission has been detected and located, the emergence of stand-off sensors and unmanned systems technology offers several opportunities. First, these systems can be layered to cover a large geographical area. GMTI radar, for instance, can detect movement in dense terrain. In the next layer, electro-optical sensors and UAS offer thermal imaging for scanning the area of interest.[65] These capabilities provide great advantages, as they require the employment of a limited number of personnel and increase the operating speed, both in terms of response time and maneuverability, to rapidly cover a large geographical area.

Similar dynamics apply to cyberspace, which is subject to the exact, rapid, continuous, and unpredictable changes as the EMS. Because of this rate of change, new challenges and opportunities emerge almost daily within cyberspace. However, only some of the challenges and opportunities will be broadly exploitable. For instance, it might be possible to have a vulnerability in a target's software application configured in a particular way. Again, if another target configures the application differently, it might not be possible to exploit the particular vulnerability in the software. Vulnerabilities are those things that can be exploited to breach confidentiality, integrity, and availability (in cyber security referred to as the CIA triad) and other security properties (e.g., non-repudiation).[66] These vulnerabilities may be found in hardware, software, and people.

A significant feature of the proliferation of these technologies is that a global, open innovation model drives them. In other words, the technologies become available to the broader public and, as a result, find their way to various actors, including non-state actors.[67] The technological advantage that the West had during the War on Terror is thus diminishing and requires the development of new tactics, techniques, and procedures (TTPs) or perhaps even the return to older ones. Indeed, there may be cases wherein special operations have to gain undetected access to the cyberspace domain by evading sophisticated Intrusion Detection Systems (IDSs) and Intrusion Prevention Systems (IPSs). Furthermore, when attempting to gain physical access, a range of additional sensors, such as closed-circuit television (CCTV), Internet of Things (IoT), and mobile communications equipment, may obstruct a clandestine approach. This is particularly important when trying to access operational areas undetected. Since these sensors themselves make use of the cyberspace domain, they must be evaded or deceived. It is important to note that such systems pose threats and provide opportunities to increase reconnaissance capability if, for example, they are "taken over."

In other cases, special operations may be conducted in densely populated areas, in plain sight, with cover identity, unplugged from modern communication devices, and applying classic encryption techniques, including traditional stenographic and cryptographic encryption schemes to avoid physical and electronic detection. This reminds us of the modes of operation employed by the men and women of the Special Operations Executive (SOE) during the Second World War (WWII) or stay-behind networks that were active during the Cold War.[68] Such modes of operation, however, are characterized by more significant differences in command and control than most Western militaries are presently accustomed to, including the limited capability to communicate quickly and continuously between forces in the field and higher Headquarters (HQ).[69] As a result, special operations need to be planned, prepared, and executed in such a way as to secure long endurance, unassisted but supported by mission command.

Most importantly, the proliferation of technology and its impact on special operations requires a new perspective on the selection, training, and equipment of the women and men conducting these missions, as they are unlikely to be recruited from the existing ranks. While this has been a long-standing tradition in most Western SOF units and provided them with exceptional tactical skill and experience, the modern surveillance environment that characterizes most of the urban OE, including city streets, train stations, and airports, offers a unique dataset from which individuals can be identified, tagged, and tracked the moment they enter a country, or after actions on the objective.[70] Therefore, having specially selected, trained, and equipped personnel capable of conducting the complete set of special operations, historically a distinctive feature of Western SOF, has become obsolete. Instead, militaries may have to look into the prospect of "single-use SOF," in which non-traditional screening processes produce ad-hoc teams trained, staffed, and equipped for a single mission. This proposition would undoubtedly bring substantial administrative and cultural challenges for organizations that pride themselves on their continued distinctiveness and their unit traditions. However, such a prospect may mitigate many of the challenges to traditional missions focused on operational preparation of the environment (OPE) or SR and may offer new opportunities to execute these missions with people more carefully tailored to a specific target or environment.

Next to the debate about the feasibility of keeping a covert or clandestine profile, David Kilcullen argues that nation-states such as Russia and China have adapted their tactics in such a way that allows them to manipulate "the threshold of detectability."[71] In his book *The Dragons and the Snakes*, Kilcullen defines this tactic as "liminal warfare" that aims to:

> Manipulate their signature so as to only pop up into the ambiguous zone of operations long enough to achieve very specific short-term goals and then to drop back down into the sub-liminal environment before we can respond. It is about manipulating their own signature, it's about creative ambiguity and it's about time – operating in the blur of the "gray zone" and surging rapidly to achieve key objectives and quickly getting back below the threshold of response before we can react.[72]

The annexation of Crimea (see Chapter 6), during which the unidentified "little green men" (i.e., soldiers without insignia on their green uniforms) seized control of key terrain features on the peninsula, illustrates an example of "liminal warfare." The cyberspace domain is a fertile ground for this tactic. Indeed, cyber espionage, subversion, and even sabotage may preserve a covert or clandestine character because the design of the Internet provides actors with the ability to maintain plausible deniability or use implausible deniability by mimicking others, creating and exploiting misinformation and disinformation, or manipulating physical and digital signatures.[73] As a result, attribution in cyberspace remains a significant challenge in today's strategic environment.

Relative superiority

From a military perspective, competitive advantage refers to a force that is superior to its opponent. Superiority may be achieved in two distinct ways: absolutely or relatively. This work argues that relative superiority will allow for mission success and remains a constant factor in special operations warfare. However, the potential ways of achieving this condition must be aligned to meet the challenges and opportunities presented by today's much more complex OE and the changes in the conduct of modern warfare.

Clausewitz introduced the concept of relative superiority. Describing the idea, he argued that "strategy decides the time when, the place where, and the forces with which, the engagement is to be fought, and through this threefold activity exerts considerable influence on its outcome."[74] These three characteristics are the operational factors of time, space, and force, which determine whether or not a mission's objective can be achieved within the limits of the available support and force structure. In determining the time and place of deploying forces,

Clausewitz explained that there are "numerous possibilities, each of which will have a different effect on the outcome of the engagement."[75] Almost two centuries later, his theory still considerably influences military strategy and special operations.

The application of the three operational factors themselves has not changed as the complexity and interconnectedness of the OE increases. However, the connotation of the areas to which these factors apply has changed dramatically since Clausewitz's time, mainly due to recent technological developments, including the emergence of cyberspace. This new domain allows operating at electronic speed and also globally, implying that effects generated in the physical dimension range from seconds to hours. In contrast, the generation of effects in the virtual dimension may occur at or near the speed of light. Furthermore, the generation of effects in the cognitive dimension has the greatest time span as they range from seconds to months or even longer. Indeed, influencing a target audience's thoughts, perceptions, emotions, and will may be achieved instantly or require a significant amount of time. Moreover, cross-dimensional effects are typically generated in a staged manner and should, therefore, be considered first, second, and third-order effects. The ability to conduct special operations in, from, or through cyberspace significantly impacts strategy and tactics, and offers an entirely new perspective on the mission's outcome.

According to Clausewitz, achieving a favorable outcome in an engagement, when stripped of all internal and external factors of influence, depends on numerical superiority, both in tactics and strategy.[76] He asserts that strategy's first principle is "that as many troops as possible should be brought into the engagement at the decisive point."[77] However, achieving absolute superiority in the three physical domains, air, land, and sea, might be infeasible due to circumstances that prevent the formation of adequate numbers of troops. Achieving absolute superiority in cyberspace might also be infeasible due to the structure of the domain. Indeed, the public nature of the domain, its low cost of gaining access, and rapid and continuous development, to mention but a few, are examples of why dominating this highly complex domain (i.e., achieving absolute superiority in specific systems or networks) is an extremely difficult, if not an impossible task, to accomplish. Clausewitz describes that if achieving absolute superiority is infeasible, "the forces available must be employed with such skill that even in the absence of absolute superiority, relative superiority is attained at the decisive point."[78] He argued that the interplay between the operational factors of time and space is critical to achieving relative superiority because, in strategy, these two factors "cover practically everything concerning the use of force."[79] In the Clausewitzian formula, the concept of relative superiority thus emphasizes that:

> The skillful concentration of superior strength at the decisive point, is much more frequently based on the correct appraisal of this decisive point, on suitable planning from the start, which leads to appropriate disposition of the forces, and on the resolution needed to sacrifice nonessentials for the sake of essentials.[80]

Reflecting on the key characteristics of special operations discussed in Chapter 1, including the usually limited size of the forces employed, achieving absolute superiority is impossible. Therefore, the concept of relative superiority, with its attention to the correct assessment of the decisive point and the detailed calculation of, especially, the operational factors of time and space, is crucial to the success of special operations.

The recognition that the concept of relative superiority is critical to the success of special operations has led to the development of a theory of special operations that builds upon the Clausewitzian conception. This theory of special operations was developed by Admiral (Ret.) William McRaven, who defined relative superiority as "a condition that exists when an attacking force, generally smaller, gains a decisive advantage over a larger, well-defended enemy."[81] This condition is not an assurance of mission success but rather an enabling condition. His theory explains the potential ways of achieving this condition while planning, preparing, and executing special operations. Its added value, therefore, lies in the "ability to illustrate which positive forces influence the success of a mission and to show how the frictions of war affect the achievement of the goal."[82] The theory was developed explicitly for direct special operations and was applied to historical case studies characterized by special operations in the traditional air, land, and sea domains. Moreover, for the special operations selected for his work, McRaven defined mission success in terms of effects generated in the physical dimension of the OE.

For developing the theoretical framework in this book, the achievement of relative superiority as a critical condition for mission success is considered a constant. However, the potential ways of achieving relative superiority need to be aligned with the ability to conduct special operations across the five domains of military operations and the EMS to meet the challenges and opportunities of the strategic environment. Indeed, the complex OE and changes in the conduct of war will require changes in the calculation of time, space, and force as the most crucial factors to achieve relative superiority across multiple domains of military operations. Moreover, the changes in the conduct of warfare call for identifying the principles of special operations that can be controlled and will influence the achievement of relative superiority. Determining effective ways to achieve and sustain relative superiority across domains will allow for the adjustment of the planning, preparation, and execution of special operations accordingly. Moreover, since this work covers the two main types of special operations, McRaven's theory is revisited and extrapolated to the twenty-first century to describe how relative superiority is achieved across domains and how this condition will enable the generation of cross-dimensional effects. In other words, when it comes to tactical performance, this book determines, discussing several different cases, what a decisive advantage across domains looks like, which principles of special operations come into play, and how these insights can be applied to the strategic model of ends, ways, and means.

To illustrate how relative superiority may have been achieved across domains, each case study will include a detailed description of the theory tailored to the particular case. The description will include positive and negative factors and key events that affected the achievement of relative superiority during mission execution. The relationship between the achievement of superiority and time deserves special attention. The longer it takes to achieve relative superiority, the greater the risk to the force of being detected, attributed, and exposed by the enemy. The element of risk is known as the area of vulnerability. Reflecting on the typical short duration of decisive action, relative superiority may be quickly achieved directly before achieving the objective. Such missions are, therefore, characterized by a relatively brief period of vulnerability. By contrast, the long duration of a shaping phase or protracted special operations indicates that achieving relative superiority may require significant time. Consequently, these missions are characterized by a more extended period of vulnerability. This distinction between the two main types of special operations (i.e., direct and indirect) requires the identification and application of the principles of special operations in such a way as to contribute positively to the achievement of relative superiority.

The principles of war

The principles of war have always served as guidelines for military operations. Various influential military thinkers have offered valuable descriptions of the general rules, guides, and principles of war. For example, Sun Tzu argued that the principles guiding military operations are deception, initiative, intelligence, leadership, logistics, maneuver, and morale.[83] Furthermore, Niccolò Machiavelli described what he defined as the general rules for military discipline, which are centered on the principles of discipline, importance of reserve forces, intelligence, morale, objective, security, surprise, use of the terrain, and ability to understand oneself as well as the adversary.[84]

Antoine-Henri Jomini, however, was the first to view warfare as a science rather than an art.[85] In one of his writings, Jomini introduced a small set of principles (i.e., mass forces against decisive points, maneuvering to engage fractions of the hostile army, and engaging at the proper times and with energy) to guide military commanders in planning, preparing, and executing military operations.[86] According to Jomini:

> The fundamental principles upon which rest all good combinations of war have always existed, and to them all others should be referred for the purpose of arriving at their respective merits. These principles are unchangeable; they are independent of the arms employed, of times, and of places.[87]

Jomini's argument is significant in that he asserts that the fundamental principles are unchangeable and independent of where and when a battle is fought and the ways and means employed.

After World War I (WWI), the British Army, greatly inspired by John Fuller, introduced its first official set of eight principles of war: Concentration, Cooperation, Economy of Force, Maintenance of the Objective, Mobility, Offensive Action, Security, and Surprise.[88] Initially, the US War Department listed nine principles in the 1921 Training Regulation 10–5, eight of which were identical to those of the

British. Still, it did not formalize the principles in US military doctrine until 1949.[89] Today, the Chinese military lists four principles, the US nine, and the Russian ten, which are all significantly different in definition and application.[90]

The study of applying the principles of war has been extended to more specialized forms of warfare, including the information-based Revolution of Military Affairs (RMA) and special operations.[91] While developing his theory, McRaven initially analyzed selected case studies for inclusion in his monograph through the lens of the US Army's eight principles of war (i.e., economy of force, maneuver, mass, offensive, security, simplicity, surprise, and unity of command) as listed in the Doctrine for Joint Operations.[92] While working on the case studies, McRaven found that some principles had to be eliminated or modified to better match the special operations context better. Accordingly, he presented the following six principles of special operations in his book: simplicity, security, repetition, surprise, speed, and purpose.[93] By applying these principles to the planning, preparation, and execution of special operations, McRaven indicated that the friction in war could be controlled and even reduced to a manageable level.[94] Finally, the most recent official document describing the principles of war is the US Doctrine on Joint Operations.[95] This doctrine lists the US Army's nine principles of war and adds three principles: legitimacy, perseverance, and restraint.[96] The US DoD added these three principles because they are relevant to how the US military uses combat power across the range of military operations.[97]

Reflecting on the rich history of the principles of warfare, some interesting observations can be made. First, there is an ongoing debate about the number of principles that should apply to warfare. Second, although there appears to be consensus about such a set of principles, thoughts on the definition and application of the principles remain pretty diverse.[98] In his article "No More Principles of War," Russel Glenn summarizes this continued debate as follows:

> History reveals that the principles of war have frequently been the subject of long and often inspired debate; their character, number, and definition have changed repeatedly. They took their present form in US Army doctrine only 49 years ago. On the one hand, this span is but a fraction of the years spent in their study. On the other hand, much has transpired since 1949. One may legitimately ask whether the principles as they stand could meet the needs of US armed forces half a century from now.[99]

Glenn's question is very much applicable to this work as well. More specifically, we should ask whether the principles of special operations identified by McRaven almost three decades ago meet the challenges and opportunities of the strategic environment. Furthermore, given that McRaven's work was focused on direct special operations, we should also ask ourselves to what extent his principles apply to indirect special operations. Indeed, having the ability to operate in all domains of military operations to achieve cross-dimensional effects while remaining undetected and unattributed for extensive periods might require the selection, definition, and application of other principles of special operations adapted to the new conditions of the strategic environment.

Notes

1 Martin van Creveld, *A History of Strategy: From Sun Tzu to William S. Lind* (Castalia House, 2015), 123–124.
2 Carl von Clausewitz, *On War*, ed. and trans. Michael Howard, Peter Paret, and Bernard Brodie (Princeton, N.J.: Princeton University Press, 1984), 593.
3 Henry A. Kissinger, *World Order* (New York: Penguin Press, 2014), 344.
4 Sohaib Athar, "Helicopter hovering above Abbottabad at 1AM (is a rare event)," Tweet, 1 May 2011 (12:58 a.m.).
5 Athar, "Helicopter hovering above Abbottabad."
6 Sohaib Athar, "A huge window shaking bang here in Abbottabad Cantt. I hope its not the start of something nasty," Tweet, 1 May 2011 (1:09 a.m.).
7 Robert Johnson, "Predicting the Future Operating Environment," *Parameters* 44, no. 1 (2014): 65–76.
8 US Department of the Army, "*Concept for Special Operations 2028*", *Pamphlet 71-20-4* (September 2020).
9 Edward Smith, "Effects Based Operations: Applying Network Centric Warfare in Peace, Crisis, and War," Office of the Assistant Secretary of Defense (OASD), Command & Control Research Program (CCRP), Washington, DC, 2003, Jacob Kipp and Texas Engineering Experiment Station, Center for Strategic Technology, *Soviet Views on Military Operations in Space* (College Station, Tex.: Center for Strategic Technology, Texas Engineering Experiment Station of the Texas A&M University System, 1986); Alfred W. Price, "The evolution of electronic warfare equipment and techniques in the USA, 1901 to 1945," Loughborough University, Doctoral Thesis, 1985.
10 Robert Citino, *Blitzkrieg to Desert Storm: the evolution of operational warfare* (Lawrence, Kan: University Press of Kansas, 2004); Pavel Tsygankov, 'Gibridnyye Voyny: ponyatiya, interpretatsii i real'nost" ['Hybrid Wars: Definitions, Interpretations and Reality'], in *'Gibridnyye Voyny' v khaotiziruyushchemsya mire XXI veka* ['Hybrid Wars' in the Chaotic World of the 21st Century], Pavel Tsygankov (ed. (Moscow: Moscow University Press, 2015); Xiao Tianliang, eds., *The Science of Military Strategy* (Beijing, National Defense University Press, 2020).
11 John Arquilla and David Ronfeldt, *ed.*, *In Athena's Camp: Preparing for Conflict in the Information Age* (Santa Monica, CA: RAND Corporation, 1997).
12 US Senate Committee on Armed Services, "Hearing to receive testimony on United States Special Operations Command's efforts to sustain the readiness of special operations forces and transform the force for future security challenges," Stenographic Transcript (April 27, 2022).
13 John Arquilla, *Bitskrieg: The New Challenge of Cyberwarfare* (Cambridge and Medford: Polity Press, 2021); Richard Clarke, quoted in Paul Mcleary, "SOCOM Likely To Boost Cyber, EW," *Breaking Defense*, May 18, 2021; Catherine Theohary, "Defense Primer: Information Operations," *Congressional Research Service* IF10771, December 1, 2021.
14 Christopher Marsh, James Kiras, and Patricia Blocksome, *ed. Special Operations Forces in an Era of Great Power Competition* (Boulder, USA: Lynne Rienner Publishers, 2019); Anthony Bovino, "Special operations forces unilateral conduct of cyberspace operations" (Master's thesis, US Naval Postgraduate School, 2020); Ryan Worrell, "EW and cyber convergence: beyond information warfare" (Master's thesis, US Air University, 2020); Arquilla, *Bitskrieg*.
15 Christian Brose, *The Kill Chain: Defending America in the Future of High-Tech Warfare*, 1st ed. (New York: Hachette Books, 2020).
16 North Atlantic Treaty Organization, "Multi-Domains Operations Conference – What We Are Learning," Allied Command Transformation, April 8, 2022.
17 Rear Adm. Hugh Howard III, quoted in "United States Senate Committee on Armed Services," April 27, 2022, 35.
18 Howard, quoted in "United States Senate Committee," 35.

19 Lt. Gen. James Slife, quoted in "United States Senate Committee," 9.
20 US Joint Chiefs of Staff, *Special Operations Review Group. Rescue Mission Report*, (Washington, DC: Joint Chiefs of Staff, 1980).
21 US Congress, "H.R.3622 - Goldwater-Nichols Department of Defense Reorganization Act of 1986," December 9, 1986.
22 Lt. Gen. James Slife, quoted in "United States Senate Committee," 9.
23 Robert D. Kaplan, "A New Cold War Has Begun," *Foreign Policy*, January 7, 2019.
24 Robert Gilpin, "Hegemonic War and International Change," in *Conflict After the Cold War: Arguments on Causes of War and Peace*, 4th ed., ed. Richard K. Betts (New York: Routledge, 2016).
25 Gilpin, "Hegemonic War."
26 Aaron L. Friedberg, "The Future of US-China Relations," *International Security* 30, no. 2 (2005): 7–45.
27 Janice G. Stein, "How Not to Think Like a Hegemon," *International Affairs* 98, no. 5 (2022): 1615–1633.
28 Richard K. Betts, "Culture: Customs, Norms, and Learning," in *Conflict After the Cold War: Arguments on Causes of War and Peace*, 4th ed., ed. Richard K. Betts (New York: Routledge, 2016).
29 Friedberg, "The Future of US-China Relations."
30 Stein, "How Not to Think Like a Hegemon."
31 Department of the Army, *The US Army in Multi-Domain Operations 2028, TP 525-3-1* (Ft. Eustis, VA: Training and Doctrine Command, 2018).
32 Tianliang, *ed., The Science of Military Strategy.*
33 Dimitry Adamsky, "Cross-domain Coercion: the Current Russian Art of Strategy," Institut Francais des Relations Internationales, Proliferation Papers 54 (2015).
34 See: William H. McRaven, *Spec Ops Case Studies in Special Operations Warfare* (New York: The Random House Publishing Group, 1995) for a detailed account of both operations.
35 See, for example, Jonathan M. House, Towards Combined Arms Warfare: A Survey of 20th-Century Tactics, Doctrine, and Organization, US Combat Studies Institute, Research Survey 2 (1984); The State Council Information Office of the People's Republic of China, "China's Military Strategy"; Gerasimov, "The Value of Science."
36 NATO Command and Control Centre of Excellence (C2COE), "Multi-Domain Command and Control: Providing a Working Description of the Term Multi-Domain C2 (MDC2)," November 2021.
37 US Air Force and US Space Force, *The Department of the Air Force role in Joint All-Domain Operations* (Washington DC: US GPO, 2021), 4.
38 House, 2.
39 Ibid.
40 House, 1–3; The State Council Information Office of the People's Republic of China, "China's Military Strategy"; Department of the Army, *The US Army in Multi-Domain Operations 2028.*
41 US Department of the Army, *The US Army in Multi-Domain Operations 2028*; Ministry of Defence, Multi-Domain Integration (MDI), Joint Concept Note (JCN) 1/20 (Shrivenham: Development, Concepts and Doctrine Centre, 2022); Jamie Shea, "Cyberspace as a Domain of Operations: What Is NATO's Vision and Strategy?," *MCU Journal* 9, no. 2 (2018); The State Council Information Office of the People's Republic of China, "China's Military Strategy."
42 US Air Force and US Space Force, "The Department of the Air Force Role in Joint All-Domain Operations," 4.
43 Larry D. Welch, "Cyberspace – The Fifth Operational Domain", Institute for Defense Analysis, (November 11 2004), 2–7.
44 US Department of Defense, *Joint Publication 3-12 Cyberspace Operations*, June 8, 2018.

45 US Department of Defense, I-2.
46 US Department of Defense, *Joint Publication 3-85 Joint Electromagnetic Spectrum Operations*, May 22, 2020.
47 US Department of Defense, v.
48 US Department of the Army, *The US Army in Multi-Domain Operations 2028*, IV-1-IV.
49 Lt. Gen. Jonathan Braga, quoted in "United States Senate Committee," 36.
50 Rear Adm. Hugh Howard III, quoted in "United States Senate Committee," 35.
51 Zhang Yuliang, *ed.*, *The Science of Military Campaigns*, trans. (Beijing: National Defense University Press, 2006), 221.
52 John Carlin, *Dawn of the Code War: America's Battle Against Russia, China, and the Rising Global Cyber Threat* (New York, NY: Public Affairs, 2019).
53 Colin Clarke and Ariane Tabatabai, "The US Designated the Revolutionary Guards as a Terrorist Group. What Happens Next?" RAND Blog, April 11, 2019.
54 Jon Henley, "Visual Guide: How Dutch Intelligence Thwarted a Russian Hacking Operation', *The Guardian*, October 4, (2018).
55 US Africa Command, "New Evidence of Russian Aircraft Active in Libyan Airspace," June 18, 2020.
56 US Department of the Navy, *Spetsnaz*, FMFRP 3-201 (Washington DC: Department of the Navy, January 1991; Igor Sutyagin and Justin Bronk, "Russia's New Ground Forces: Capabilities, Limitations and Implications for International Security," RUSI Whitehall Papers, June 28, 2017; Department of the Army, *Concept for Special Operations 2028.*
57 Ronan Bergman, "The Dubai Job," GQ, January 4, 2011; U.K. Foreign & Commonwealth Office, "Evidence of Russia's Involvement in Salisbury Attack," Speech delivered at UN Security Council, September 6, 2018.
58 Giorgos Gaitanakis, George Limnaios, and Konstantinos Zikidis, "On the use of AESA (Active Electronically Scanned Array) Radar and IRST (InfraRed Search&Track) System to Detect and Track Low Observable Threats," *MATEC Web of Conferences* 304 (2019): 1–8.
59 John Entzminger, Charles Fowler and William Kenneally, "JointSTARS and GMTI: past, present and future," *EEE Transactions on Aerospace and Electronic Systems* 35, no. 2 (1999): 748–761.
60 John A. Richards, "GMTI Radar Minimum Detectable Velocity," Sandia National Laboratories, April 2011.
61 US Department of the Navy, "Acoustic Rapid Commercial Off-the-Shelf Insertion (A-RCI) for AN/BQQ-10(V) Sonar," FY17 Navy Programs, A-RCI 9, 2017: 137–138.
62 US Special Operations Command (USSOCOM), "Science and Technology - Preparing for the Future 2020-2030 Special Operations Forces Small Unit Dominance (SOF SUD)," Tampa (2018).
63 Bryan Clark, Whitney McNamara, and Timothy Walton, "Winning the Invisible War: Gaining an Enduring US Advantage in the Electromagnetic Spectrum," (Washington, DC: Center for Strategic and Budgetary Assessments, 2019).
64 Peter Roberts, "Electronic Warfare and Cumulative Risk," Western Way of War, podcast, series 2, episode 29, 17 December 2020.
65 Jack Watling, "Sharpening the Dagger: Optimising Special Forces for Future Conflict," RUSI Whitehall Papers (2021).
66 Suhail Qadir, S. M. K. Quadri, "Information Availability: An Insight into the Most Important Attribute of Information Security," *Journal of Information Security* 7, no. 3 (2016): 185–194.
67 Audrey K. Cronin, "Technology and Strategic Surprise: Adapting to an Era of Open Innovation," *Parameters* 50, 3 (2020): 71–84.
68 Kevin D. Stringer, "Survival in the Russian Occupied Zone Command and Organization in Resistance Underground Operations," *Military Review* (2021): 125–132.
69 Trent Lythgoe, "Beyond Auftragstaktik the Case Against Hyper-Decentralized Command," *Joint Forces Quarterly* 96 (2020).

70 Ilya Arkhipov and Jane Rudnitsky, "In Moscow, Big Brother Is Watching and Recognizing Protesters," Bloomberg, May 2, 2021; Chris Buckley and Paul Mozur, 'How China Uses High-Tech Surveillance to Subdue Minorities', *New York Times*, May 22, 2019.
71 David Kilcullen, *The Dragons and the Snakes: How the Rest Learned to Fight the West* (New York NY: Oxford University Press, 2020), 119.
72 Kilcullen, "Liminal and Conceptual Envelopment: Warfare in the Age of Dragons," *Small Wars Journal*, May 26, (2020).
73 Rory Cormac, Richard J. Aldrich, Grey is the new black: covert action and implausible deniability, *International Affairs* 94, no. 3 (2018): 477–494.
74 Clausewitz, 194.
75 Ibid.
76 Clausewitz, *On War*.
77 Ibid., 195.
78 Ibid., 196.
79 Ibid., 196.
80 Ibid., 197.
81 McRaven, 4.
82 Ibid.
83 Sun Tzu. *The Art of War*, trans. by Samuel B. Griffith (Oxford University Press, 1963).
84 Niccolo Machiavelli, *The Art of War*, with introduction by Neal Wood (New York: Da Capo Press, 1990).
85 Antoine Henri Jomini, Jomini and His Summary of The Art of War, in *Roots of Strategy Book 2, Three Military Classics*, ed. Brig. Gen. J. D. Little (Harrisburg, PA: Stackpole Books, 1987).
86 John I. Alger, *The Quest for Victory* (Westport, CT: Greenwood Press, 1982).
87 Marshall l. Fallwell, "The Principles of War and the Solution of Military Problems," *Military Review* 35 (1955): 50.
88 Alger, *The Quest for Victory*.
89 G. K. Carmichael, "Principles of War and Their Application to Strategy and Tactics," *Naval War College Review* 3, no. 8 (1950): 23–42.
90 Fallwell, "The Principles of War"; Tianliang, ed., *The Science of Military Strategy*.
91 Leonard G. Litton, "The Information-Based RMA and the Principles of War," *Air & Space Power Chronicles* (2000); McRaven, *Spec Ops*.
92 McRaven, *Spec Ops*; US Department of Defense, *Joint Publication 3-0 Joint Operations* (Washington DC: Department of Defense, January 2017.
93 McRaven, *Spec Ops*.
94 Ibid.
95 US Department of Defense, *Joint Publication 3-0*.
96 Ibid.
97 Ibid.
98 Fallwell, "The Principles of War."
99 Russell W. Glenn, "No More Principles of War," *Parameters* 28, no. 1 (Spring 1998): 64.

Bibliography

Adamsky, Dimitry. "Cross-Domain Coercion: the Current Russian Art of Strategy," Institut Francais des Relations Internationales, *Proliferation Papers* 54 (2015), https://www.ifri.org/sites/default/files/atoms/files/pp54adamsky.pdf.

Alger, John I. *The Quest for Victory* (Westport, CT: Greenwood Press, 1982).

Arkhipov, Ilya, and Jane Rudnitsky. "In Moscow, Big Brother Is Watching and Recognizing Protesters," *Bloomberg*, May 2, 2021, https://www.bloomberg.com/news/articles/2021-05-02/in-moscow-big-brother-is-watching-and-recognizing-protesters#xj4y7vzkg.

Arquilla, John. *Bitskrieg: The New Challenge of Cyberwarfare* (Cambridge and Medford: Polity Press, 2021).
Arquilla, John, and David Ronfeldt. eds., *In Athena's Camp: Preparing for Conflict in the Information Age* (Santa Monica, CA: RAND Corporation, 1997).
Athar, Sohaib. "Helicopter hovering above Abbottabad at 1AM (is a rare event)," Tweet, 1 May 2011 (12:58 a.m.), https://twitter.com/reallyvirtual/status/64780730286358528.
Athar, Sohaib. "A huge window shaking bang here in Abbottabad Cantt. I hope its not the start of something nasty: -S," Tweet, 1 May 2011 (1:09 a.m.), https://twitter.com/reallyvirtual/status/64783440226168832?lang=zh-Hant.
Bachman, Charles. "Provisional Model of Open System Architecture," *ACM SIGCOMM Computer Communication Review* 8, no. 3 (1978): 49–61.
Bergman, Ronan. "The Dubai Job," GQ, January 4, 2011, https://www.gq.com/story/the-dubai-job-mossad-assassination-hamas.
Betts, Richard K. "Culture: Customs, Norms, and Learning," in *Conflict After the Cold War: Arguments on Causes of War and Peace*, 4th ed., Richard K. Betts, ed. (New York: Routledge, 2016).
Betts, Richard K. "International Realism: Anarchy and Power," in *Conflict After the Cold War: Arguments on Causes of War and Peace*, 4th ed., Richard K. Betts, ed. (New York: Routledge, 2016).
Bovino, Anthony. "Special operations forces unilateral conduct of cyberspace operations" (Master's thesis, US Naval Postgraduate School, 2020), https://calhoun.nps.edu/handle/10945/66591.
Brose, Christian. *The Kill Chain: Defending America in the Future of High-Tech Warfare*, 1st ed. (New York: Hachette Books, 2020);
Buckley, Chris, and Paul Mozur. "How China Uses High-Tech Surveillance to Subdue Minorities", *New York Times*, May 22, 2019, https://www.nytimes.com/2019/05/22/world/asia/china-surveillance-xinjiang.html.
Carlin, John. *Dawn of the Code War: America's Battle Against Russia, China, and the Rising Global Cyber Threat* (New York, NY: Public Affairs, 2019).
Carmichael, G.K. "Principles of War and Their Application to Strategy and Tactics," *Naval War College Review* 3, no. 8 (1950): 23–42.
Citino, Robert. *Blitzkrieg to Desert Storm: the Evolution of Operational Warfare* (Lawrence, Kan: University Press of Kansas, 2004).
Clark, Bryan, Whitney McNamara, and Timothy Walton. *Winning the Invisible War: Gaining an Enduring US Advantage in the Electromagnetic Spectrum* (Washington, DC: Center for Strategic and Budgetary Assessments, 2019), https://csbaonline.org/uploads/documents/Winning_the_Invisible_War_WEB.pdf.
Clarke, Colin and Tabatabai, Ariane. "The US Designated the Revolutionary Guards as a Terrorist Group. What Happens Next?', RAND Blog, 11 April 2019, https://www.rand.org/blog/2019/04/the-us-designated-the-revolutionary-guards-as-a-terrorist.html.
Clarke, Richard quoted in Matthew Cox. "Less Door-Kicking, More Influencing: The Changing Role of Special Operators," *Military.com*, May 12, 2020, https://www.military.com/daily-news/2020/05/12/less-door-kicking-more-influencing-changing-role-special-operators.html.
Clarke, Richard quoted in Paul Mcleary. "SOCOM Likely to Boost Cyber, EW," *Breaking Defense*, May 18, 2021, https://breakingdefense.com/2021/05/socom-likely-to-boost-cyber-ew/.
Clausewitz, von Carl. *On War*, ed. and trans. by Michael Howard, Peter Paret, and Bernard Brodie (Princeton, N.J.: Princeton University Press, 1984).
Cormac, Roy, and Richard J. Aldrich. "Grey Is the New Black: Covert Action and Implausible Deniability," *International Affairs* 94, no. 3 (2018): 477–494.
Creveld, van Martin. *A History of Strategy: From Sun Tzu to William S* (Lind (Castalia House, 2015).

Cronin, Audrey K. "Technology and Strategic Surprise: Adapting to an Era of Open Innovation," *Parameters* 50, no. 3 (2020): 71–84.
Entzminger, John, Charles Fowler, and William Kenneally. "JointSTARS and GMTI: Past, Present and Future," *EEE Transactions on Aerospace and Electronic Systems* 35, no. 2 (1999): 748–761.
Fallwell, Marshall I. "The Principles of War and the Solution of Military Problems," *Military Review* 35 (1955).
Friedberg, Aaron L. "The Future of US-China Relations," *International Security* 30, no. 2 (2005): 7–45.
Gaitanakis, Giorgos, George Limnaios, and Konstantinos Zikidis. "On the Use of AESA (Active Electronically Scanned Array) Radar and IRST (InfraRed Search&Track) System to Detect and Track Low Observable Threats," *MATEC Web of Conferences* 304 (2019): 1–8.
Gilpin, Robert. "Hegemonic War and International Change," in *Conflict After the Cold War: Arguments on Causes of War and Peace*, 4th ed., Richard K. Betts, ed. (New York: Routledge, 2016).
Gray, Colin S. *Explorations in Strategy* (Westport: CT, Praeger, 1998).
Glenn, Russell W. "No More Principles of War," *Parameters* 28, no. 1 (1998): 48–66.
Henley, Jon. "Visual Guide: How Dutch Intelligence Thwarted a Russian Hacking Operation", *The Guardian*, October 4, 2018, https://www.theguardian.com/world/2018/oct/04/visual-guide-how-dutch-intelligence-thwarted-a-russian-hacking-operation.
House, Jonathan M. Towards Combined Arms Warfare: A Survey of 20th-Century Tactics, Doctrine, and Organization, US Combat Studies Institute, Research Survey 2 (1984).
Johnson, Robert. "Predicting the Future Operating Environment," *Parameters* 44, no. 1 (2014): 65–76.
Jomini, Antoine Henri. Jomini and His Summary of The Art of War, in *Roots of Strategy Book 2, Three Military Classics*, Brig. Gen. J. D. Little, ed. (Harrisburg, PA: Stackpole Books, 1987).
Kaplan, Robert. "A New Cold War Has Begun", *Foreign Policy*, January 7, 2019, https://foreignpolicy.com/2019/01/07/a-new-cold-war-has-begun/.
Kilcullen, David. *The Dragons and the Snakes: How the Rest Learned to Fight the West* (New York NY: Oxford University Press, 2020).
Kilcullen, David. "Liminal and Conceptual Envelopment: Warfare in the Age of Dragons," *Small Wars Journal*, May 26, 2020, https://smallwarsjournal.com/jrnl/art/liminal-and-conceptual-envelopment-warfare-age-dragons.
Kipp, Jacob W. Texas Engineering Experiment Station *Soviet Views on Military Operations in Space* (College Station, Tex: Center for Strategic Technology, Texas Engineering Experiment Station of the Texas A&M University System, 1986).
Kissinger, Henry A. *World Order* (New York: Penguin Press, 2014).
Lawrence, Paul, and Jay Lorsch. "Differentiation and Integration in Complex Organizations," *Administrative Science Quarterly* 12, no. 1 (1967): 1–47.
Hart, Liddell Basil. *Strategy of the Indirect Approach* (London: Faber and Faber Limited, 1941).
Litton, Leonard G. "The Information-Based RMA and the Principles of War," *Air & Space Power Chronicles* (Sept. 2000).
Lythgoe, Trent. "Beyond Auftragstaktik: The Case Against Hyper-Decentralized Command," *Joint Forces Quarterly* 96 (2020).
Machiavelli, Niccolo. *The Art of War*, with introduction by Neal Wood (New York: Da Capo Press, 1990).
Marsh, Christopher, James Kiras, and Patricia Blocksome. eds., *Special Operations Forces in an Era of Great Power Competition* (Boulder, USA: Lynne Rienner Publishers, 2019).
McRaven, William H. *Spec Ops Case Studies in Special Operations Warfare* (New York: The Random House Publishing Group, 1995).

Ministry of Defence, Multi-Domain Integration (MDI). Joint Concept Note (JCN) 1/20 (Shrivenham: Development, Concepts and Doctrine Centre, 2022), https://assets.publishing.service.gov.uk/government/uploads/system/uploads/attachment_data/file/950789/20201112-JCN_1_20_MDI.PDF.

NATO Command and Control Centre of Excellence (C2COE). "Multi-Domain Command and Control: Providing a Working Description of the Term Multi-Domain C2 (MDC2)," November 2021, https://c2coe.org/wpcontent/uploads/Library%20Documents/Study/20210430%20MDC2%20Long%201.0.pdf.

North Atlantic Treaty Organization. "Multi-Domains Operations Conference – What We Are Learning," Allied Command Transformation, April 8, 2022. https://www.act.nato.int/articles/multi-domains-operations-lessons-learned.

Price, Alfred. "The evolution of electronic warfare equipment and techniques in the USA, 1901 to 1945," Loughborough University, Doctoral Thesis, 1985, https://hdl.handle.net/2134/7410.

Qadir, Suhail, and S.M.K. Quadri. "Information Availability: An Insight into the Most Important Attribute of Information Security," *Journal of Information Security* 7, no. 3 (2016): 185–194.

Richards, John A. "GMTI Radar Minimum Detectable Velocity," Sandia National Laboratories, April 2011, https://www.osti.gov/servlets/purl/1011708.

Roberts, Peter. "Electronic Warfare and Cumulative Risk," Western Way of War, podcast, series 2, episode 29, December 17, 2020, https://rusi.org/multimedia/electronic-warfare-and-cumulative-risk.

Shea, Jamie. "Cyberspace as a Domain of Operations: What Is NATO's Vision and Strategy?" *MCU Journal* 9 (2018): 2.

Smith, Edward. "Effects Based Operations: Applying Network Centric Warfare in Peace, Crisis, and War," Office of the Assistant Secretary of Defense (OASD), Command & Control Research Program (CCRP), Washington, DC, 2003), https://apps.dtic.mil/sti/pdfs/ADA457292.pdf.

Stein, Janice G. "How Not to Think Like a Hegemon," *International Affairs* 98, no. 5 (2022): 1615–1633.

Stringer, Kevin D. "Survival in the Russian Occupied Zone Command and Organization in Resistance Underground Operations," *Military Review* (2021): 125–132.

Sun, Tzu. *The Art of War*, trans. by Samuel B. Griffith (Oxford University Press, 1963).

Sutyagin, Igor, and Justin Bronk. "Russia's New Ground Forces: Capabilities, Limitations and Implications for International Security," RUSI Whitehall Papers, June 28, 2017.

Tianliang, Xiao. ed., *The Science of Military Strategy* (Beijing: National Defense University Press, 2020).

Theohary, Catherine. "Defense Primer: Information Operations," *Congressional Research Service* IF10771, December 1, 2021, https://crsreports.congress.gov/product/details?prodcode=IF10771.

The State Council Information Office of the People's Republic of China. "China's Military Strategy," trans., (Beijing, May 2015), https://china.usc.edu/prc-state-council-chinas-military-strategy-2015-may-26-2015.

Tsygankov, Pavel. "Gibridnyye Voyny: Ponyatiya, Interpretatsii i real'nost" ['Hybrid Wars: Definitions, Interpretations and Reality'], in *'Gibridnyye Voyny' v Khaotiziruyushchemsya Mire XXI Veka* ['Hybrid Wars' in the Chaotic World of the 21st Century], Pavel Tsygankov, ed. (Moscow: Moscow University Press, 2015).

U.K. Foreign & Commonwealth Office. "Evidence of Russia's Involvement in Salisbury Attack," Speech delivered at UN Security Council, September 6, 2018, https://www.gov.uk/government/speeches/you-dont-recruit-an-arsonist-to-put-out-a-fire-you-especially-dont-do-that-when-the-fire-is-one-they-caused.

US Africa Command. "New Evidence of Russian Aircraft Active in Libyan Airspace,"June 18, 2020, https://www.africom.mil/pressrelease/32941/new-evidence-of-russian-aircraft-active-in-li.

US Air Force and US Space Force. *The Department of the Air Force Role in Joint All-Domain Operations* (Washington DC: US GPO, 2021), https://www.doctrine.af.mil/Portals/61/documents/AFDP_3-99/AFDP%203-99%20DAF%20role%20in%20JADO.pdf.

US Department of Defense. *Joint Publication 3-0 Joint Operations* (Washington DC: Department of Defense, January 2017), https://irp.fas.org/doddir/dod/jp3_0.pdf.

US Department of Defense, *Joint Publication 3-12 Cyberspace Operations*, June 8, 2018, https://info.publicintelligence.net/JCS-CyberspaceOperations.pdf.

US Department of Defense. *Joint Publication 3-85 Joint Electromagnetic Spectrum Operations*, May 22, 2020, https://www.jcs.mil/Portals/36/Documents/Doctrine/pubs/jp3_85.pdf.

US Department of the Army. *The US Army in Multi-Domain Operations 2028, TP 525-3-1* (Ft. Eustis, VA: Training and Doctrine Command, 2018), https://adminpubs.tradoc.army.mil/pamphlets/TP525-3-1.pdf.

US Department of the Army. *"Concept for Special Operations 2028", Pamphlet 71-20-4* (September 2020), https://apps.dtic.mil/sti/pdfs/AD1128559.pdf.

US Department of the Navy. *Spetsnaz*, FMFRP 3-201 (Washington DC: Department of the Navy, January 1991), https://ia800705.us.archive.org/22/items/MManuals/UsMarines-SpetsnazRussianSpecialForces1991.pdf.

US Department of the Navy. "Acoustic Rapid Commercial Off-the-Shelf Insertion (A-RCI) for AN/BQQ-10(V) Sonar," FY17 Navy Programs, A-RCI 9, 2017: 137–138, https://www.dote.osd.mil/Portals/97/pub/reports/FY2017/navy/2017arci.pdf?ver=2019-08-19-113710-477.

US Joint Chiefs of Staff. *Special Operations Review Group. Rescue Mission Report* (Washington, DC: Joint Chiefs of Staff, 1980), https://nsarchive2.gwu.edu/NSAEBB/NSAEBB63/doc8.pdf.

US Senate Committee on Armed Services. "Hearing to receive testimony on United States Special Operations Command's efforts to sustain the readiness of special operations forces and transform the force for future security challenges," Stenographic Transcript (April 27, 2022), 1–65. https://www.armed-services.senate.gov/imo/media/doc/22-21_04-05-2022.pdf.

US Special Operations Command (USSOCOM). "Science and Technology – Preparing for the Future 2020-2030 Special Operations Forces Small Unit Dominance (SOF SUD)," Tampa (2018), https://www.socom.mil/st/STPublicDocLibrary/PROVIDE%20SOLUTIONS%20-%20ST%20SOF%20Small%20Unit%20Dominance.pdf.

Watling, Jack. *Sharpening the Dagger: Optimising Special Forces for Future Conflict* (RUSI Whitehall Papers, 2021).

Welch, Larry D. "Cyberspace – The Fifth Operational Domain," Institute for Defense Analysis, November 11, 2004, 2–7, https://www.ida.org/-/media/feature/publications/2/20/2011-cyberspace---the-fifth-operational-domain/2011-cyberspace---the-fifth-operational-domain.ashx.

Worrell, Ryan. "EW and cyber convergence: beyond information warfare" (Master's thesis, US Air University, 2020), https://apps.dtic.mil/sti/pdfs/AD1112357.pdf.

Yuliang, Zhang. ed., *The Science of Military Campaigns* (Beijing: National Defense University Press, 2006).

Part II

Case Studies in Direct Special Operations

3 The Russian Raid on Hostomel Airport

Introduction

Russia successfully annexed the Crimean Peninsula in March 2014, as described in Chapter 6 of this book. Shortly after that, Russia started to actively support pro-Russian separatists with the occupation of the Luhansk and Donetsk Oblasts, which led to the war in eastern Ukraine. After the failure of the Minsk agreements signed in September 2014 (Minsk Protocol) and February 2015 (Minsk II), which aimed to end the war in Donbas, Ukraine moved further toward the West's sphere of influence and openly shared its ambition to become a member of the European Union (EU) and the North Atlantic Treaty Organization (NATO).

With the election of Volodymyr Zelensky as President of Ukraine on March 31, 2019, the move toward the West became even more outspoken and threatened Russia's national interest of maintaining influence and power over former Soviet states in its "Near Abroad."[1] This threat persuaded Russian President Vladimir Putin to adjust his strategic objectives from controlling the Donbas to overthrowing the Zelensky administration and installing a pro-Russian government equal to the one in Crimea and Donbas.

Background

In the spring of 2021, US intelligence agencies informed senior US government officials about the first indications of a further escalation of the ongoing Russo-Ukraine war. By then, the Russian military had begun significantly building up forces along Ukraine's borders.[2] The purpose of this build-up was twofold: first, it allowed Russia to preposition sufficient military force to be employed in case Putin would decide to invade Ukraine. Second, overtly moving military forces and equipment allowed Russia's senior leaders to assess the international community's responses.[3] At the same time, Microsoft observed Russian Advanced Persistent Threat (APT) actors operationally preparing the cyberspace domain. Although the level of coordination between the various APTs remains questionable, Microsoft concluded that the Russian activities aimed to secure persistent access in Ukrainian systems to collect intelligence on defense and foreign partnerships and to prepare for future cyberattacks on critical infrastructure in support of an invasion.[4]

DOI: 10.4324/9781003365112-5

The US intelligence community did not assess that Putin was planning for the invasion that spring.[5] The two main arguments supporting the assessment were the absence of critical enablers that typically deploy with the Battalion Tactical Groups (BTGs) and the lack of signs that Russia's special services were shaping the information environment.[6] As observed in other cases, including Georgia (2008) and Crimea (2014), the Kremlin announced that the build-up of military forces was part of a mobilization exercise.[7] At that time, General Paul Nakasone, director of the National Security Agency (NSA) and commander of US Cyber Command (USCYBERCOM), confirmed that the troop movements were part of an exercise.[8]

The key takeaway for Russia's senior leaders from that mobilization exercise was that they could assemble the required military forces faster than the West could provide Ukraine with military capabilities.[9] This important lesson seems to have boosted Putin's confidence that he could invade Ukraine without the risk of Western intervention.[10] Russia's decade of military modernization further increased Putin's confidence. The architect of this modernization process, Chief of the General Staff of the Russian Armed Forces, General Valery Gerasimov, held a strong conviction that Russia's military was among the most powerful in the world.[11] Russian forces would, therefore, be able to defeat the numerically inferior Ukrainian military quickly and decisively.

Around September 2021, US intelligence detected new Russian troop build-ups. American analysts observed that the size and scale of this build-up, as well as the disposition and composition of the Russian forces, differed from what they had observed during the previous years. Additionally, the Central Intelligence Agency (CIA) had been picking up indications that Russia was planning a full-scale invasion of Ukraine. By October, Nakasone was convinced that Russia was going to invade Ukraine.[12]

In the following weeks, the US government made several efforts to de-escalate the situation. For example, CIA director Bill Burns went to Moscow and spoke to Putin and his closest advisers. Furthermore, during a G20 summit in Rome, US President Joe Biden shared the intelligence about Russia's intentions with British Prime Minister Boris Johnson, French President Emmanuel Macron, and German *Bundeskanzler* Angela Merkel. Secretary of State Anthony Blinken also briefed Zelensky on the American intelligence at the Glasgow Summit on November 2. Finally, the Biden administration took an unusual step by declassifying and releasing parts of the intelligence to the public. By doing so, the US exposed to the world Russia's preparations along the Ukrainian border.[13]

In parallel to exposing Russia's preparations, the US and its allies and partners strengthened Ukraine's military capabilities. For example, Western Special Operations Forces (SOF) have been in the country since 2014 and provided training to Ukrainian SOF.[14] Furthermore, Nakasone deployed a "hunt forward team" to Kyiv in the first week of December to strengthen Ukrainian digital networks.[15]

In addition to the military assistance effort, the West prepared a package of sanctions that would impose significant costs on Russia in case Putin would decide to invade Ukraine. This deterrence by punishment approach emphasized and convinced Putin that there was a limited risk of Western military intervention. Classic deterrence theory describes two distinct types of deterrence. Deterrence by denial emphasizes discouraging an aggressor by articulating the unlikelihood of a successful outcome.[16] Deterrence by punishment, on the other hand, is deterring aggression by threatening penalties with severe consequences.[17] Hence, deterrence by punishment is a more indirect approach than deterrence by denial, as the severity of the punishment should make an attack very costly.[18] Deterrence by denial, however, is a more reliable form than deterrence by punishment.[19]

In Ukraine's case, the West announced it would focus its military efforts on strengthening NATO's eastern flank rather than Ukraine.[20] Former US president Barack Obama already stated in 2016 that Ukraine might be a vital national interest for Russia but not for the US He added that Ukraine is not a NATO member state and hence is vulnerable to Russian aggression. The US, he concluded, should "be very clear about what our core interests are and what we are willing to go to war for."[21] Subsequently, US President Biden further stated that US forces would not be part of defensive efforts against Russian aggression.[22] Instead, the West threatened to punish Russian aggression by imposing economic sanctions.[23] This threat fits the deterrence by punishment approach, as it would raise the costs of an invasion for Russia. The goal of the sanctions was twofold: first, they had to have a maximum impact on Russia and its ability and willingness to continue the war effort. Second, the sanctions needed to have a limited effect on US allies and partners. To this end, the US coordinated closely with the EU, United Kingdom (UK), Canada, and Japan, among others. That coordination continued throughout December 2021 and January 2022 to be optimally prepared.[24]

The fundamental tenet of direct versus extended deterrence explains the failure of this approach. Direct deterrence aims to prevent an attack on one's own territory, while extended deterrence involves an attack on third parties.[25] Ukraine is not a member of NATO nor protected by bilateral security guarantees, such as the US-Taiwan Relations Act.[26] Furthermore, the West assumingly was unwilling to intervene, fearing nuclear retaliation.[27] Finally, when the US decided to withdraw its military personnel from Ukraine, it signaled a comforting message to Russia regarding the risk of encountering US or NATO forces in case of an invasion.[28] Hence, only the deterrent of economic sanctions was just not powerful enough, and by January 2022, it was probably too late to change the course of events.[29]

Russia developed its invasion strategy on three key assumptions. First, speed. Moving fast would be critical in rendering any potential Western response irrelevant. Second, decapitating the Ukrainian government would be vital to gain popular support from the pro-Russian population in Ukraine for the invasion, subsequent occupation, and installment of a pro-Russian government. Russian intelligence anticipated that Ukrainian senior leaders, including Zelensky, would either flee the country, get arrested, or be killed. Russia further estimated that potential civil unrest would be insignificant, as the invasion's speed would shock Ukraine's society and hinder mobilization efforts. If demonstrations occur, elements of the Russian National Guard (*Rosgvardia*) would tail the BTGs to conduct riot control. To prevent escalation, Russia would deploy the Federal Security Service (*Federal'naya sluzhba bezopasnosti Rossiyskoy Federatsii*, or FSB) throughout the country with the task of intimidating, arresting, or even assassinating local Ukrainians who posed a threat to Russia's objectives. Finally, seizing critical infrastructure such as nuclear power plants and financial institutions would allow Russia to control civil society by shutting down electricity, heating systems, and access to financial resources. In sum, a classical *fait accompli* was Russia's strategy of choice to achieve its strategic objectives quickly.

The Russian military developed a strategy trying to achieve three strategic effects. First, precise air and missile strikes had to degrade Ukraine's key air and maritime defense capabilities and disrupt its command-and-control structures. The Russians would not target critical infrastructure such as the power grid, railways, and major roads, as they had planned to use them during the operation themselves. Second, Russian ground forces had to fix the considerable number of Ukrainian combat forces deployed along the contact line in the Donbas region. Third, the FSB and Main Intelligence Directorate (*Glavnoye Razvedyvatelnoye Upravlenie*, or GRU), Russian SOF (*sily spetsial'nykh operatsii*, or SSO), and elite airborne troops (*Vozdushno-desantnye voyska Rossii*, or VDV) had to degrade Ukraine's ability and willingness to resist by toppling its government, as well as seizing critical infrastructure such as the nuclear power plants, financial institutions, the parliament, and airfields.

The purpose of seizing nuclear power plants was threefold. First, the sites would provide the Russian military with a protective area from where they could perform key warfighting functions such as command-and-control, as the chances of Ukrainian attacks would be extremely limited due to the significant risk of a nuclear incident. Second, nuclear power plants generate more than 50–55 percent of Ukraine's electricity.[30] Gaining control of them would allow Russia to use energy as a weapon against civil society by shutting down electricity and heating systems. Finally, by seizing the installations, Russia could signal the risk of a nuclear fallout to the West if, against all expectations, they would consider intervening militarily.

The ways to achieve the three strategic effects reveal a clear contradiction regarding the risk of exposure. On the one hand, planning and preparation for the first two strategic effects required strict operational security measures, while the success of the execution depended on speed. Effectively degrading Ukrainian resistance, on the other hand, meant that Russia's special services had to actively engage the Ukrainian population in the months prior to the invasion, which increased the risk of exposing the operation. To this end, Russia effectively delayed the talks surrounding the Minsk Accords while simultaneously deploying its special services and senior military personnel to eastern Ukraine to shape the environment, prepare for the invasion, and gain popular support from the Ukrainian population.

The special services successfully gained widespread support in Donbas due to the significant differences between the urban population and those living in the countryside. The former is very progressive and pro-EU and played an active role in the Revolution of Dignity, popularly known as the Maidan Revolution, which took place in February 2014 in central Kyiv. The latter, by contrast, are less educated and still organized according to former communist structures, as these remote areas are rich in heavy industry (steel) and coal mines, but at the same time lack basic amenities such as running water and sewage systems. As a result, the social contract with the people in the countryside is weaker than the ties between urban inhabitants and the capital. Russian operatives, therefore, successfully recruited pro-Russian separatists consisting of the foremen of the various labor unions that could swiftly mobilize large numbers of people to occupy administrative centers, police stations, and the like. Also, old principles such as loyalty to the communist system coupled with an old patronage system favored Russia's preparation of the information environment.

In July 2021, Ukrainian intelligence began detecting some Russian preparations. During that month, Russia's special services started collecting survey data from the Ukrainian population to better understand their perception of the country's political, economic, and military situation. The results of those surveys indicated that most of the Ukrainian people had little trust in their political leadership and were primarily concerned with the poor economic situation of their country. Most Ukrainians also believed that further escalation of the ongoing war with Russia was unlikely.[31] Finally, at that time, approval rates for Zelensky were below 30 percent. At the same time, Putin stated in an essay that he believed most Ukrainians had a most favorable attitude toward Russia, as they were, in his view, part of a shared civilization. The separation of both countries, according to Putin, was the result of historical political mistakes. The Ukrainian government, however, could restore the relationship between the two countries as soon as they would stop acting as a puppet of the West.[32] Of note: After the collapse of the Soviet Union in 1991, Ukraine organized a referendum on the Act of Declaration of Independence on December 1, 1991. The outcome of the referendum could not have been clearer, as 92.3 percent of the voters favored the declaration of independence made by the *Verkhovna Rada* of Ukraine (Supreme Council of Ukraine) on August 24, 1991.[33]

Ukrainian intelligence detected no further indications of Russian shaping activities to prepare for the invasion. Drawing from historical successes such as Budapest (1956), Prague (1968), Kabul (1979), and more recently, Crimea (2014), it is likely that Russia's political and military leaders were over-optimistic of the "strategic shock effect" that a full-scale invasion would have on Ukraine's ability and willingness to resist. These assumptions may have led to the decision to launch the operation without having thoroughly prepared the information environment.

The Russian offensive included three main axes of advance, as shown in Figure 3.1. First, on the southern axis, units from the Southern Military District Command Post would start their advance from Crimea. Recognizable by the tactical symbol "Z in a square," the units would first gain control over the North Crimean Water Canal and then advance into the direction of Enerhodar Dnipro Power Line Crossing, Zaporizhzhia Nuclear Power Plant, and Zaporizhzhia Thermal Power Station. Additionally, the forces would seize the bridges over the Dnipro, which would allow the Russians to further advance into the direction of Kherson and Mykolaiv, and Voznesensk, a small city but strategically significant due to its proximity to the South Ukrainian Nuclear Power Plant. Finally, the units on the southern axis would advance eastwards and surround Mariupol, situated at the Sea of Azov and home to the Azovstal Iron and Steel Works. After the initial advance, a naval

Figure 3.1 Hostomel Airport relative to the three Russian main axes of advance.

landing force attached to Russia's Black Sea Fleet and positioned along Ukraine's southern coast would complement the ground forces on this axis. These additional forces would seize Odesa, Ukraine's third most populous city and a major seaport.

Units from the Western Military District Command Post would advance into Ukraine from Kursk, Belgorod, and Voronezh. Russian strategic deception, also known as *maskirovka*, played a key role on this axis. The Russians successfully made Ukrainian intelligence believe that their main axis of advance would be coming from the Donbas region in eastern Ukraine instead of from the direction of Gomel in neighboring Belarus. Due to this successful Russian deception operation, Ukraine further enforced its combat forces on the Eastern Front. Having successfully deceived Ukrainian intelligence, the Russian units on this axis would advance deep into Ukrainian territory to fix the Ukrainian forces deployed along the contact line in the Donbas region. If successful, the Russians would be able to cut off a substantial part of Ukraine's combat power, leaving Kyiv's northern flank unprotected.

Finally, units from the Eastern District Military Command Post were to exploit the successful deception operation and would start their advance from Gomel in Belarus and Bryansk in Russia. These units, recognizable by the tactical symbol "V" (Gomel axis) and "O" (Bryansk axis), were the main effort and tasked with the encirclement and subsequent occupation of Kyiv. One of the key targets of the forces on the main axis was to seize Hostomel Airport, which lies only ten kilometers outside the capital. The airport played a decisive role in the opening days of the invasion.

Hostomel airport

Hostomel Airport, also known as Antonov Airport, was built in 1959. The airport is situated in Hostomel, a small town just outside Kyiv. The airport is a major international cargo hub, home to the Antonov State Enterprise, the Ukrainian state-owned Aerospace and Air Defense Company. The airport is known for hosting the Antonov An-225 Mriya, which is considered the world's largest airplane. Next to its cargo function, the airport is in use by the Ukrainian Air Force.

The airport is known for its long runway, historically used to test cargo planes. The length of the runway combined with its proximity to the Ukrainian capital illustrates its strategic significance during an armed conflict, particularly when an invading force wants to use it as a bridgehead to insert personnel and equipment to seize Kyiv quickly.

The strategic significance of Hostomel Airport became public when CIA Director Burns visited Ukraine in January 2022. During his visit, Burns informed Ukrainian government officials that US intelligence indicated that in the event of an invasion, Russia would attempt to seize Hostomel Airport and use it as an air-head to bring in additional troops to overthrow the Ukrainian government. According to US intelligence, this swift action would cause chaos among the Ukrainian senior leadership, significantly weaken its armed forces, and allow Russia to install a puppet government. Despite the risk of mass protests among the Ukrainian population against the Russian occupation, US intelligence assessed this to be a highly likely scenario through which the Russian leadership believed they could be successful.[34]

Planning and preparation

Little is known about the planning and preparation of the raid on Hostomel Airport. However, we know that the Russian High Command had assigned their elite airborne troops (VDV) to conduct the mission. The VDV probably conducted their planning and preparation for the raid from airbases in southern Belarus, as American commercial satellite company Maxar released images of Russian Su-25 ground-attack aircraft and attack helicopters at Luninets airfield in Belarus on February 18, 2022.[35] The airbase is approximately 50 kilometers north of Ukraine's border, making it suitable for helicopter insertion. Another possibility was the Bolshoi Bokov Airbase situated in southern Belarus, as Maxar photographed more than 50 Russian helicopters as well as troops on February 25.[36] This airbase is located approximately 100 km from Kyiv, which made it also very suitable for the operation.

Because the Russian planning for the invasion was already broadly exposed, the Russians needed to conceal the date and time of the operation. Documents allegedly found by Ukrainian forces indicate that the Russians had planned for combat operations between February 20 and March 16.[37] According to US senior officials, Putin may not have determined the precise date of the invasion well in advance but rather decided to launch it when he thought the time was ripe.[38] As a result, the

Russian High Command may have ordered the VDV to conduct the raid on short notice, giving them limited time to make detailed preparations or conduct a full-dressed rehearsal. A rare Russian perspective on the operation, given by a captured soldier of the VDV, seems to confirm the limited preparation time available. In an interview with journalist Volodymyr Zolkin, the soldier stated that the briefing that the tactical units had received prior to the operation was of poor quality.[39] Other sources assert that the VDV had three days to prepare for the raid, which is a relatively short time for proper preparation. Perhaps the Russians were trying to decrease the likelihood that the use of modern telecommunication and social media could expose the plans. The truth may never be known, but the case findings indicate that protecting the date and time of a high-risk operation to exploit the element of surprise comes at a cost.

The raid

On February 23, 2022, a day before the raid on Hostomel Airport, Unit 74455 of the GRU deployed malware against roughly 300 Ukrainian targets, including the energy, financial, IT, and public sectors.[40] Microsoft detected the attack, immediately alerted the Ukrainian government, and assisted them in stopping it.[41] The next morning (February 24), at 4:00 a.m., Russia started to attack Ukraine with missiles and artillery fire. This preliminary attack was not fully accurate, as the Russians could not destroy most of Ukraine's defenses, including air-defense systems. Russia was able, however, to destroy two specific air-defense sites responsible for screening the Dnipro River north of Kyiv. By doing so, the Russians could open a gap in the Ukrainian air defenses, which enabled them to penetrate Ukrainian airspace from Belarus. In conjunction with the preliminary attack, Microsoft observed a variety of Russian APTs conducting offensive cyberoperations in support of the Russian objectives.[42] The cyberattack launched against the Viasat satellite communications network, used by the Ukrainian military and police force, was considered to be the most prominent. The effectiveness of the attack was substantial, as Ukrainian command-and-control was disrupted for several days. Then, at around 5:30 a.m., President Putin officially announced a "special military operation" to "demilitarize and denazify" Ukraine.[43]

Only minutes after Putin had made his announcement, Lieutenant Colonel Vitaliy Rudenko, a battalion commander in the Ukrainian National Guard and responsible for the defense of Hostomel Airport, received a phone call from the Ukrainian High Command informing him to prepare for battle. Shortly after the phone call, a cruise missile landed on the parade ground of Rudenko's barracks, and half an hour later, multiple cruise missiles struck the airport itself. By now, it was clear to Rudenko that the situation in Ukraine was escalating, but he did not believe there was an immediate threat to the airport itself. Despite his reservations, Rudenko and his troops decided to occupy their defensive positions and wait for whatever would follow. At the same time, the airport's civilian employees were arriving and discussing what to do with all the civilian airplanes, including the Antonov An-225 Mriya. They discussed several options and eventually decided to spread the fleet across the airport in case of further missile attacks.

At approximately 9:30 a.m., a helicopter assault force consisting of between 30 and 40 helicopters from the Russian Air Force took off from staging areas across southern Belarus. A mix of Mi-24P Hind-F and Kamov Ka-52 "Alligator" Hokum-B attack helicopters led the formation. Following the course of the Dnipro River in the southern direction, the helicopters flew at extremely low altitudes to evade detection. However, the Ukrainian National Guard, which was responsible for the defense of a hydroelectric plant located at the Kyiv Cistern, spotted the approaching helicopters and immediately engaged them with 9K38 Igla man-portable air-defense systems (MANPADS). One of the Ka-52s was hit and crashed into the Dnipro River. The helicopters immediately broke out of formation and dodged in a Western direction, heading toward Hostomel Airport.

Meanwhile, Rudenko and his troops were still in their defensive positions. Around 10:30 am, they heard the sound of rotor blades but could not see the helicopters because they were flying extremely low to the ground. When the first Ka-52s appeared above the trees, they immediately engaged the Ukrainian defensive positions with guided missiles, rockets, and 30mm cannons. In doing so, the Russians were able to destroy some parts of the airport's defenses but failed to weaken them significantly. The Ukrainians engaged the Russian helicopters with their personal weapons as well as MANPADS. The Russian pilots immediately deployed their flares when engaged, but this did not prevent the Ukrainians from downing multiple helicopters with their MANPADS. The approximately 80 civilian employees who were still at the airport quickly fled to the bomb shelter under the cafeteria while others hid in the sewers.

To sustain the air attack, the Russians had established two forward arming and refueling points located in fields near the city of Mazyr in southern Belarus. At approximately one hundred kilometers north of Hostomel Airport, the two sites allowed the Russians to refuel and re-arm their attack helicopters quickly.

After the initial air attack, 18 multipurpose Mil Mi-8AMTSh Hip-Cs 20 medium-lift utility helicopters approached the airport. On board the Mi-8s were an estimated three hundred airborne troops of the VDV, likely belonging to the 11th and 31st Guards Air Assault Brigade. Once at their designated landing sites, the VDV began to seize the airport. Rudenko's battalion consisted of only 120 personnel and was thus significantly smaller than the Russian landing force. He quickly realized that they were unable to stop the VDV despite the preparations his unit had made in response to the CIA director's warning, which they had received the previous month. By applying the principle of speed, the VDV could overwhelm Rudenko's troops and successfully capture most of the target by the end of the morning.

Recognizing the superiority of the VDV and by now low on ammunition, Rudenko ordered his troops to withdraw from the airport, but not before making a last effort to block the runway. The withdrawal was chaotic, as some of the troops used personal vehicles to flee while the majority climbed the airport's outer fences and ran into the built-up area in its vicinity. Not all Rudenko's troops could escape, as the VDV captured several of them. In the afternoon, CNN reporter Matthew Chance and his crew broadcasted the events live from the airport's outer fence. He also spoke to a VDV commander who was responsible for the security at one of the gates; he informed Chance that the Russians had taken control of the airport.[44]

The VDV then started to prepare for the next phase of the operation, namely the Tactical Air Landing Operation (TALO), to allow for approximately 18 Ilyushin Il-76 cargo planes from Russia's Military Transport Aviation Command to land at the airport. Based at Kresty Airbase in Pskov, Russia, these airplanes were already airborne, carrying an additional 1,000 troops and heavy material from the 76th Guards Air Assault Division. After Rudenko's order, however, his forces cleverly blocked the runway with inoperable heavy equipment, tractors, and fuel trucks. It was one of the few preparations made in the days prior to the invasion but turned out to play a decisive role in the defense of Kyiv, as the lightly equipped Russian forces were unable to remove the heavy equipment in a timely manner, delaying the TALO and forcing the IL-76s to remain airborne in a holding pattern.

While the VDV seized Hostomel Airport, they could not benefit from the advantage gained through the successful strategic deception operation described at the beginning of this chapter. The VDV was unable to exploit their advantage because Ukrainian forces had arrived in an assembly area just south of Kharkiv on the night of February 23, waiting for an escort to their position along the contact line at the Eastern Front. As a result, Russian units from the Western Military District Command Post that would fix Ukrainian forces in the Donbas region found themselves in a meeting engagement with these Ukrainian units on the early morning of February 24.

By now fully aware of the significant threat posed by the potential of a Russian airhead at Hostomel Airport, General Valerii Zaluzhnyi, Commander-in-Chief of the Armed Forces of Ukraine, ordered the 72nd mechanized brigade to counterattack the airport with heavy artillery. The artillery barrage further destroyed the landing strip and denied the IL-76s from landing, thereby necessitating them to fly back to Kresty Airbase. In addition to the 72nd mechanized brigade, Zaluzhnyi directed combat units from the 80th Air Assault Brigade, 4th brigade of the Ukrainian National Guard, SOF, and the Georgian Legion (a group of battle-hardened foreigners) to join the counterattack. The Ukrainians could direct artillery fire on the VDV by establishing observation posts around the airport. With the TALO canceled, the lightly equipped VDV lacked the necessary support and resources to offset the Ukrainian counterattack, so they withdrew to the northern part of the airport.

When night fell, both sides had established defensive positions at the airport. The Ukrainians made several attempts to further counterattack VDV positions, but these came quickly to a halt because of Russian direct fires resulting in several wounded personnel. Because the Ukrainians considered making further advancements under the cover of darkness to be unfeasible, they decided to remain static for the remainder of the night. The next morning (February 25), reports of Russian armored columns advancing over the Gomel axis came in. These ground forces, organized into BTGs, would encircle Kyiv and breach its outer defenses. The Russians could gain a tactical advantage over the Ukrainians by employing electronic warfare capabilities, including jamming. Simultaneously, Russian APTs launched additional cyberattacks on Ukrainian systems, 90 percent of which took place in Kyiv, Donetsk, and Luhansk Oblasts.[45] Whether these cyberattacks occurred in close coordination with ground maneuvers remains doubtful, as only a few preceded an attack by Russian ground forces.[46]

The Ukrainian military was able to delay the Russian advance because they had removed all road signs in the previous weeks. They were also jamming navigation apps, and the Russians were using old Soviet maps that were outdated and did not reflect the actual environment, as various villages, for example, could not be identified on the map. Upon entering these villages, Russian personnel would try to contact the local population to find out their exact location. Once the Russians left, the local villagers then reported the positions of these forces to the Ukrainian military, who would then quickly attack them with artillery fire or Unmanned Aerial Systems (UAS). Additionally, the Ukrainians had destroyed their own infrastructure, such as bridges and the Kozarovychi dam across the Irpin River, 30 km northeast of Hostomel Airport, to manipulate the terrain and hamper the Russian advance. In doing so, the Ukrainians were able to flood the Irpin floodplain, making the terrain highly unsuitable for maneuver warfare. As a result, the Russian ground forces had to advance slowly and often in single file, which made them extremely vulnerable to artillery fires and flank attacks.

Despite these Ukrainian countermeasures, the Russian ground forces successfully linked up with the VDV forces that were still hiding in the northern part of the airport on the morning of February 25. Jointly, and further complemented with additional VDV troops inserted by helicopters, they reattacked the airport as well as the surrounding villages of Bucha, Irpin, and Hostomel. Recognizing the scale of this second Russian assault, the Ukrainian High Command decided to withdraw its forces from the airport and have them reinforce defensive positions at the outskirts of Kyiv. These positions offered significantly more protection than the ones previously occupied at the airport and still allowed them to engage the Russian forces that would soon fully occupy the airport.

Around noon, the Russians had successfully retaken complete control of the airport. Because the Ukrainian defenses in the surrounding area received additional combat units and reserve units (mobilized on February 24), the Russians could not penetrate them. Furthermore, Ukrainian forces were able to continuously attack the Russian forces at the airport with artillery fire. Consequently, the Russians could not try to repair the damaged runway, let alone create an area from where they could employ critical warfighting functions, such as logistics and command-and-control, to support their efforts to seize Kyiv. These dynamics further delayed the Russian ground assault. In addition, because the Russians could not establish a suitable logistical hub close enough to the front but out of reach of Ukrainian artillery, problematic situations arose when it came to supplying their forward-deployed combat units. Due to these factors, coupled with weak Russian leadership, poorly trained forces, and lack of communication devices, a 65-kilometer convoy consisting of Russian ground forces was observed north of Kyiv. Remaining static on the main supply route, the convoy became extremely vulnerable to Ukrainian hit-and-run actions, from which it suffered severely. These events, broadcasted live by international media, turned out to be Russia's "Culminating Point," a concept described by Carl von Clausewitz as the point in time where a defensive force can stop an adversarial offensive before the attacker has achieved its objectives.[47]

In sum, Russia's failure to quickly seize Hostomel Airport prevented them from establishing the airhead required to bring in additional forces and heavy material vital to the intended *fait accompli* strategy. In the days that followed, fierce battles took place in the vicinity of the airport as well as the surrounding villages of Bucha, Irpin, and Hostomel. Ukrainian combat forces were able to inflict severe damage on Russian armored columns by using Tactics, Techniques, and Procedures (TTPs) derived from the concept of swarming. They combined them with the employment of UAS, the use of closed-circuit television (CCTV) security systems, and information on the disposition of these columns received from the local population.

Additionally, effective Ukrainian counterintelligence and information provided by personnel from Russia's special services themselves prevented several attempts to kill Zelensky. Once united in the Committee for State Security (*Komitet Gosudarstvennoy Bezopasnosti*, or KGB), the historic relationships between the old intelligence services of both countries still dictate trust and information-sharing, including with Zelensky. Due to this significant threat and the heavy fighting on the outskirts of Kyiv, the US offered to exfiltrate Zelensky and his family out of Ukraine. Zelensky rejected the offer and instead released a video from downtown Kyiv at 8:00 pm on February 24 in which he and his closest Cabinet members stated: "We are all here, defending our independence."[48] This strong message boosted Ukrainian morale and willingness to resist during the opening days of the invasion.

By February 25, the Russians had conquered territory along all three axes of advance. Their main objectives, however, of seizing Kyiv and overthrowing the Ukrainian government, had failed. Ukrainian reinforcements on the defensive lines surrounding Kyiv prevented the Russians from further advancing toward the capital. After a month of intense fighting, Russia shifted its focus to controlling the Donbas region and preparing for the battle for Mariupol and withdrew its combat forces from Kyiv.

Analysis

Policy

On the morning of February 24, President Putin announced a "special military operation" to "demilitarize and denazify" Ukraine.[49] In reality, however, Russia invaded Ukraine with the political objective "to be recognized as a great power with its own distinct sphere of influence" and "that Russia has a moral right to predominance within 'its' sphere of influence."[50] A Ukraine that is falling into the Western sphere of influence, including becoming a member of the EU or NATO, is a threat to this vital Russian national interest.[51] To secure and advance that national interest, overthrowing the pro-Western Zelensky administration and replacing it with a pro-Russian one had become Russia's political objective.[52] To achieve that objective, Russia developed a grand strategy that included employing military capabilities in conjunction with other instruments of national power.

Military strategy

Russia intended to quickly seize control of Ukraine by using a classical *fait accompli* strategy. Russia has successfully employed the concept throughout history. For example, the Soviet Union celebrated successes with this strategy in Budapest (1956), Prague (1968) and Kabul (1979). In this case, the military strategy aimed to achieve the following ends: first, preliminary fires had to destroy Ukrainian air defenses, command-and-control (C2) structures, and ammunition depots. Second, Russia's special services had to topple the Ukrainian government. Third, Russian SOF and elite airborne units had to seize several key objectives, including critical infrastructure, financial institutions, the parliament, and airfields.

The first end would be achieved by a precise missile and airstrike campaign combined with cyberattacks, while the second would be accomplished by capturing or killing Ukrainian senior leaders. The seizure of Hostomel Airport, only ten kilometers from the center of Kyiv, was vital to achieving the third end, namely seizing the financial institutions and the parliament. The airport would be captured by conducting a raid followed by a TALO. This would allow for sufficient troop numbers and heavy equipment to be brought in close enough to the center of Kyiv to occupy the financial institutions and parliament swiftly. According to a CIA assessment, these actions would cause chaos among the Ukrainian senior leadership, significantly weaken its armed forces, and, despite the risk of mass protests in Kyiv and other parts of the country, allow for the installment of a pro-Russian government.[53] Whether the Russian military strategy was inherently logical depended on its suitability, feasibility, and acceptability.

Suitability: could the Russians execute the operation with the available resources?

The VDV quickly seized the airport by applying the principles of speed and surprise. The lightly equipped forces, however, could not remove the heavy equipment that the Ukrainians had placed on the runway. As a result, the TALO could not be executed, forcing the VDV to sustain its presence on a fixed target while unsupported by ground or by air. The vulnerability of the VDV became visible when Ukrainian mechanized combat units were tasked to counterattack. Heavy artillery barrages aimed at the airport resulted in substantial casualties among the VDV and forced them to withdraw to the forested area in the northern part of the airport. Moreover, the lack of Russian air superiority prevented the VDV from inflicting any considerable damage on Ukrainian defenses and turned them into a static force that had to dig defensive positions while waiting for a linkup with Russian mechanized ground forces.

These Russian ground forces were unable to reach the airport in a timely manner because they were delayed by using old and inaccurate Soviet maps and also by Ukrainian defensive measures, including removing all road signs and jamming navigation apps. These delays presented the Ukrainians with the opportunity to attack these Russian forces and gave them precious time to reinforce the defensive line on the north side of Kyiv. The delay also allowed the Ukrainians to blow up the

Kozarovychi dam across the Irpin River. As a result, the terrain north of Kyiv was flooded, thereby making it unsuitable for Russian maneuver warfare.

In sum, although the VDV was initially able to seize Hostomel Airport, the limited size of the landing force prevented them from effectively establishing the airhead to allow for the TALO. Without additional forces and heavy equipment, the Russians could not further exploit the elements of speed and surprise, which were vital to achieving their objectives. The operation was not successful due to the completely unforeseen and unexpected maneuver by the Ukrainians that made the runway useless and illustrated the Russians' miscalculation in the operational factors of time, space, and force.

Feasibility: did achieving the objectives create the strategic effects that contributed to the desired end?

The objective of the raid was to seize Hostomel Airport and use it as an airhead to bring in additional forces and heavy equipment to further advance toward Kyiv and overthrow the Ukrainian government. According to the CIA, this swift action would cause chaos among the Ukrainian senior leadership, significantly weaken its armed forces, and allow Russia to install a puppet government. Based upon that assessment, effectively seizing the airport, which is located just on the outskirts of Kyiv, quickly followed by the TALO, would likely have given the Russians immediate access to the capital. Whether this would have enabled them to overthrow the Ukrainian government depends on many other factors, but would probably have allowed the Russians to seize key administrative centers in Kyiv. Therefore, the airport seizure was vital to the Russian strategy. Because the Russians failed to achieve this objective, they did not create the requisite strategic effects to contribute to the desired end.

Acceptability: did the strategic effects sought to justify the objectives pursued and consider tangible and intangible costs?

To determine whether the objectives (i.e., complete Russian annexation of Ukraine) met the conditions of acceptability based on consequences for Russia, it is crucial to assess two distinct forms of costs. Costs resulting from Western military support, but also those resulting from the imposed economic sanctions.

Even though the US and NATO did not intervene militarily in Ukraine, the Russian invasion has significantly unified its members, including Finland and Sweden, joining the alliance in 2023 and 2024, respectively, which is certainly not in Russia's interests. Quickly after the invasion, NATO members started supplying the Ukrainians with military equipment and ammunition, including advanced weapons systems such as the M142 HIMARS light multiple rocket launcher and MIM-104 Patriot surface-to-air missile (SAM) system. The US alone has provided more than $50 billion in assistance to Ukraine as of February 2023, with other NATO allies offering substantial aid as well.[54] This support turned out to be vital to Ukraine's ability to sustain its defensive military operations against Russia. Putin likely miscalculated Western unity and willingness to provide Ukraine with the necessary military resources despite Western European countries' dependence on Russian oil and gas.

Next to the costs of Western military support to Ukraine are the economic sanctions imposed by the West. Following the invasion of Ukraine, the US, Canada, and the EU imposed three types of economic sanctions on Russia. First, they gave Russian state-owned businesses in the financial and energy sectors, as well as the defense industry, restricted access to Western financial markets. Second, they placed an embargo on the export of high-technology oil exploration and production equipment to Russia. Finally, they restricted the export of designated military and so-called dual-use goods to Russia. These sanctions damaged the Russian economy, as the fall in oil prices resulted in significant downward pressure on the value of the ruble and increased capital flights.[55]

However, Europe's dependence on Russian oil and gas turned out to be too great to impose full sanctions. Because of that dependence, Russia's oil and gas exports remained exceedingly high and provided Russia with substantial financial resources, as the EU spent approximately €58bn on importing Russian fossil fuels during the first six months of the war.[56] Additional restrictions imposed on technology exports to Russia intended to reduce Russia's ability to sustain the war by limiting defense production.[57] These sanctions are adding to Russia's already existing problem of emigration. Its best and brightest are seeking opportunities in western-friendly states.[58]

The continuous and increased military support coupled with sustained economic sanctions did not lead to conflict termination. However, it still does prevent Russia from trying to achieve its strategic objectives. Whether the West would have responded similarly if the Russians had succeeded in the *fait accompli* is questionable. In that case, it is not unthinkable that the West would have reacted comparably as it did after the annexation of Crimea, and the tangible and intangible costs for Russia to achieve its objectives may have been acceptable. At the same time, the West's economic interests in Ukraine are far greater than in Crimea because of the substantial amounts of grain, which may have led to a strong Western response.

In sum, the inherent logic of the strategy developed by Russia proved to be invalid. The ends, ways, and means do not answer the question of suitability, feasibility, and, to a lesser extent, acceptability in a positive way. The main defect in the Russian strategy was found in the relationship between the ends and means, as the initial landing force was simply too small and too lightly equipped to secure the airport effectively and start with the execution of the TALO. As a result, the Russians lost the momentum at the beginning of the operation and were not able to regain it after that. Consequently, the objective of establishing the airhead (i.e., the third end) vital to the success of their strategy could not be achieved due to the limited resources employed (i.e., means). Finally, contrary to the assessments of various military analysts, Russia's decade of military modernization did not fulfill its promise. Additionally, experiences from Crimea, Donbas, and Syria, as well as Russia's quantitative superiority, did not enable them to execute the operation successfully. As a result, Russia did not achieve its strategic objectives and ultimately failed to advance its national interest of maintaining power and influence over former Soviet states in its "Near Abroad."

Risk

What follows is an assessment of the risks associated with the Russian strategy. This assessment includes considering the consequences of success and failure; examining the ends, ways, and means relative to the strategic environment; and determining which factors contributed to the generation of strategic effects. Since this chapter focuses on the raid on Hostomel Airport, which took place in the opening days of the war, the risk assessment is focused on the immediate consequences of that operation.

What assumptions did the Russians make in their strategy, and what was the effect of any wrong assumption?

Most importantly, Russia assumed that the *fait accompli* strategy would yield a comparable success as it had done in Hungary (1956), Czechoslovakia (1968), Afghanistan (1979), and more recently Crimea (2014). Due to overconfidence, Russia did not start the operation with massive fires, as Russian doctrine prescribes. Therefore, Ukrainian air defenses remained mainly intact, preventing the Russians from gaining air superiority. As a result, the Russian Air Force failed to neutralize the Ukrainian Air Force or guarantee close air support to Russian ground forces. These findings correspond with previous scholarship that found that a military *fait accompli* frequently fails and leads to armed conflict. Indeed, research indicates that out of 65 *faits accomplis* attempted between 1945 and 2018, only half of them were completed successfully.[59] The main reason for these failures is that the aggressor state miscalculated the redline for a new status quo.[60]

Secondly, Russia assumed that the Ukrainian population would be hospitable to the Russian military. Signals Intelligence (SIGINT) revealed that Russian forces were expecting the Ukrainian population to view them as liberators. When entering local villages while advancing over the various axes, Russian soldiers, therefore, tried to contact the local population. Rather than welcoming the Russian troops, the Ukrainian population stayed inside their houses and did not interact with them at all. Instead, they reported the positions of the Russian units to the Ukrainian military. Consequently, as soon as the Russians left the villages and entered rural areas, they quickly faced either artillery fire or Ukrainian SOF. SOF teams employed swarming tactics and UAS tailored for specific offensive action forms. For example, UAS performed top attacks on Russian tanks and other armored vehicles that turned out to be highly effective.

Finally, Russia assumed that its military was superior at the tactical and operational levels. Russia's BTGs appeared to be too weak to penetrate deep into Ukrainian territory. The BTGs encountered significant losses while trying to breach Ukrainian defenses in urban areas. Furthermore, Ukrainian forces had manipulated the terrain by blowing up bridges and dams, which made the terrain unsuitable for maneuver warfare. Consequently, armored columns stopped along the major roads leading to these urban areas and exposed their combat service support units. These factors, coupled with weak Russian leadership, poorly trained forces, and lack of communication devices, made the Russian logistical lines extremely vulnerable to Ukrainian hit-and-run actions from which they suffered severely.

Did the Russians incorporate flexibility or adaptability into their strategy? If required, how could they have modified their strategy, and at what cost?

The Russian strategy demonstrated limited flexibility. On the second day of the invasion, it became clear that Hostomel Airport had become unsuitable as an airhead. Furthermore, Russia's special services failed to kill or capture Zelensky, senior members of his Cabinet, or senior military commanders. Although some objectives were achieved on other axes of advance, mainly the southern one, the momentum was lost. Without air superiority, Russian ground forces found themselves engaged in intense battles near Chernihiv and Kharkiv. However, the few units that could breach Kyiv's defenses were destroyed by Ukrainian defenses. It was only by the beginning of April, after attrition and exhaustion made the majority of Russia's BTGs incapable of fighting, that Russia adjusted its strategy to controlling the entire Donbas region as well as preparing for the battle for Mariupol and withdrew its combat forces from around Kyiv.

How did other actors react to the Russian invasion?

On February 25, NATO Heads of State and Government issued a statement in which they strongly condemned the Russian invasion by calling it "the gravest threat to Euro-Atlantic security in decades."[61] At the same time, the alliance made clear it would not intervene militarily, as Ukraine was not a member of the alliance. Invoking Article 5, which states that an attack on one member of NATO is deemed an attack on all, was therefore not applicable.

President Putin was quick to announce that economic sanctions would be viewed as an act of war, against which he threatened nuclear escalation. Despite these threats, the West did impose economic sanctions on Russian financial institutions, state enterprises, and oligarchs. The most significant impact resulted from the sanctions imposed by the US on the Russian central bank. However, European countries did not agree to fully sanction Russian oil, as their dependence was too great.

The People's Republic of China (PRC), on the other hand, did not condemn Russian aggression but rather blamed the US for inciting the conflict, as it would serve the US's national interests. Although no evidence was found that indicates Putin informed the PRC's leadership about his plans to invade Ukraine, some scholars find it "hard to believe that Putin would have started without a green light from President Xi Jinping."[62] Notably, only 20 days before Russia launched the operation, it signed a "no limits pact" with the PRC based on pragmatic transnationalism, which granted the PRC "critical influence over Russia's decision-making."[63] At the same time, and due to their interests in Eurasia, Chinese diplomats told EU officials that Russian use of nuclear weapons would be unacceptable.[64]

How did unforeseen factors play a role in the strategy?

In their attempt to quickly seize Hostomel Airport, the Russians successfully deceived the Ukrainians, leaving Kyiv's northern flank relatively unprotected. Despite this successful deception operation, the Russians could not exploit their advantage because Ukrainian forces arrived in an assembly area just south of Kharkiv on the night of February 23, waiting for an escort to their position along the contact line at the Eastern Front. As a result, Russian units from the Western Military District Command Post that were tasked with fixing Ukrainian forces in the Donbas region found themselves in a meeting engagement with these Ukrainian units on the early morning of February 24. This might be a classic example of Clausewitzian *Friktion* or just limited Russian satellite reconnaissance capability. It is also possible that the Russians did detect these Ukrainian forces but failed to share this information with their tactical ground forces, as the orders to launch the invasion had already been given. The presence of these Ukrainian forces prevented the Russians from encircling the Ukrainian defenses and allowed the Ukrainians to send reinforcements to Kyiv when it became clear that the Russian main axis of advance was coming from the north.

Tactics

The raid on Hostomel Airport illustrates that, in the special operations context, political objectives will only be secured if they can be achieved through successful tactical action. Even though the political objective was clearly defined, it was not aligned with a supporting military strategy. This demonstrates the lack of strategic sense among the Russian policymakers and military commanders. One reason could be the specific Russian civil–military relations whereby Putin has absolute power over the military. Over the years, Putin has become more isolated, and very few people, if any, were able or dared to speak up to him as this equals disloyalty. Therefore, it is likely that the Russian military leadership was confronted with a political leader who asked them to achieve the unachievable while simultaneously denying any constructive civil–military dialogue about the feasibility of the operation. Subsequently, any concerns about the feasibility of seizing the airport by daylight that tactical-level commanders may have brought up were likely disregarded by their military superiors.

Although superior tactical action becomes meaningless when it serves an invalid strategy that does not support the political objective, it remains valuable to assess the tactical activities employed by the VDV to determine whether they gained relative superiority throughout the raid.

Did the VDV achieve relative superiority throughout the raid?

The VDV did achieve relative superiority by applying strategic deception, surprise, and speed throughout the planning, preparation, and execution of the operation. First, during the planning and preparation phases of the raid, the Russians successfully made Ukrainian intelligence believe that their main axis of advance would be coming from the Donbas region in eastern Ukraine instead of from the direction of Gomel in neighboring Belarus. As a result, the Ukrainians decided to further enforce their combat forces on the Eastern Front to the extent that approximately 60 percent of their warfighting capabilities were deployed near the Donbas region, leaving Kyiv's northern flank relatively unprotected. By doing so, the Russians could decrease the area of vulnerability and set the appropriate conditions to launch a successful surprise attack.

The VDV further decreased the area of vulnerability throughout the initial stages of the execution phase by applying the principle of surprise. The Ukrainian forces defending the airport believed it would be a target for indirect fire and thus did not expect a raid to seize it. Even when several missiles were fired on the airport and nearby barracks on the early morning of February 24, a raid was believed to be unlikely. Only when the defenders heard the sound of rotor blades, quickly followed by the air attack, did it become clear that the CIA's warning had become a reality. By surprising the airport's defenders, the Ukrainian defenses were ineffective and enabled the VDV to gain an advantage before they reached the point of vulnerability. By doing so, the VDV reduced the area of vulnerability by half. Figure 3.2 illustrates this by the point of vulnerability starting above the Z-axis (i.e., relative superiority line) and allowing the VDV to start the assault with relative superiority. Having achieved relative superiority at this early stage in the operation meant that the Russian chance of mission success exceeded the chance of mission failure.

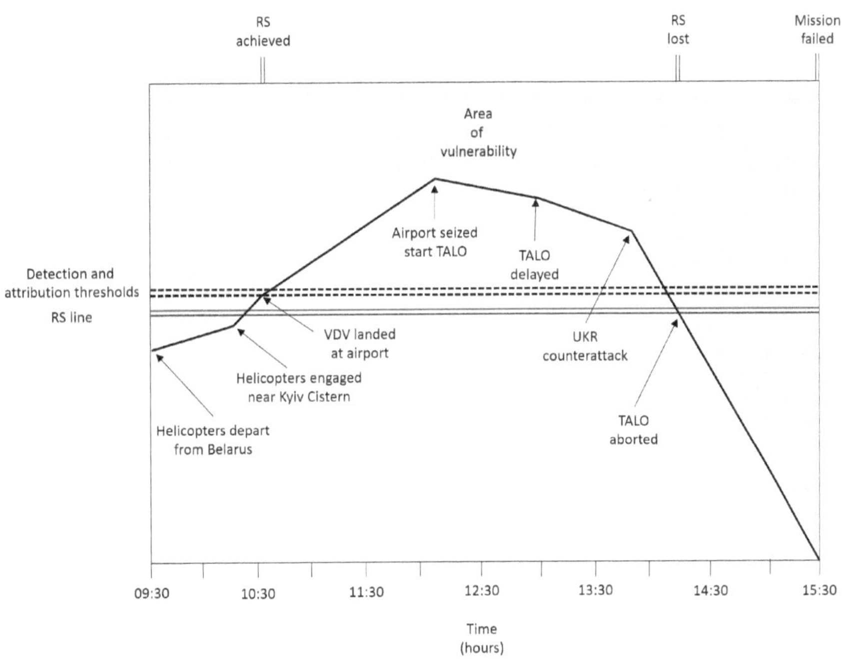

Figure 3.2 Relative superiority graph for the Russian raid on Hostomel Airport.

During the assault, the VDV could fully achieve relative superiority by applying the principle of speed. Immediately after being dropped at the various landing sites, the VDV started to seize the airport. The VDV overwhelmed the Ukrainian forces with their speed of advance, which allowed them to capture most of the airport's critical components by the end of the morning. Recognizing their inability to defend the airport, the Ukrainians withdrew into the surrounding urban areas, allowing the VDV to take complete control of the airport in the afternoon. At this point in time, the chance of mission success was even further outweighing the chance of mission failure, thereby further decreasing the area of vulnerability.

If achieved, how did the VDV sustain its relative superiority throughout the operation? If not, why did they fail?

The VDV failed to sustain their relative superiority because they miscalculated the risk that comes with the principle of economy of force. To complete the mission, the VDV had to prepare for the next phase of the operation, namely the TALO, which would allow for the IL-76 cargo planes carrying additional troops and heavy equipment to land at the airport. The VDV, however, was unable to prepare the TALO because Ukrainian forces had blocked the runway with inoperable heavy equipment such as fuel trucks and tractors. The lightly equipped VDV was unable to remove these materials in a timely manner, delaying the TALO and forcing the IL-76s to remain airborne in a holding pattern. This delay prevented the VDV from sustaining relative superiority and allowed the Ukrainian High Command to direct additional combat forces to launch a counterattack at the airport. The VDV's failure to remove the heavy equipment from the runway illustrates poor Russian planning, as one could expect such defensive actions to take place on an airport that is vulnerable to air assault and TALO. Figure 3.2 illustrates that the VDV's sustained presence at the airport resulted in a greater area of vulnerability.

When General Zaluzhnyi ordered the 72nd mechanized brigade to counterattack with heavy artillery, the landing strip was destroyed, and the TALO was canceled. With the TALO aborted, a lack of air superiority, and only lightly equipped, the VDV had limited resources at its disposal and was forced to withdraw to the forested area in the northern part of the airport. At this point in time, the VDV could not sustain relative superiority as the likelihood of mission failure began to outweigh the probability of mission completion. This is depicted in Figure 3.2 by the steep declining curve that now reaches far below the Z-axis, thereby dramatically increasing the area of vulnerability. The loss of relative superiority illustrates the vulnerability of elite units when attacking fixed targets, and a sustained presence is required. Indeed, the fact that special operations promise to be economical also comes with risk. One black spot may result in a negative disproportionate return on investment, illustrated through the principle of economy of force, which in this case led to mission failure.

Conclusion

As described in the introductory chapter of this book, political objectives are defined by policymakers, while military commanders are responsible for the employment of tactics. Exploiting the full potential of special operations is thus highly dependent on bilateral consultation and mutual understanding of strategy development and tactical behavior.

The raid on Hostomel Airport illustrates the difficulty of achieving political objectives through tactical action. Contrary to the findings presented in Chapter 6 of this book, this case study underscores the Russian policymakers and military commanders' limited level of strategic sense, as they miscalculated the conversion of tactical performance into strategic effects. The operation, therefore, illustrates one in which the Russians did not fulfill the strategic promise of special operations because they misaligned their policy, strategy, and tactics. The unhealthy Russian civil–military relationships that characterize Putin's regime may have contributed significantly to this misalignment, as the Russian military leadership was likely unable to address any possible concerns regarding the feasibility of achieving the objectives with tactical action through the chain of command.

Notes

1 "Near Abroad" is an exclusively Russian term used to denote the countries of the former Soviet Union.
2 Erin Banco, Garrett Graff, Lara Seligman, Nahal Toosi, and Alexander Ward, "Something Was Badly Wrong: When Washington Realized Russia Was Actually Invading Ukraine," *Politico*, February 24, 2023.
3 Mykhaylo Zabrodskyi, Jack Watling, Oleksandr V Danylyuk, and Nick Reynolds, "*Preliminary Lessons in Conventional Warfighting from Russia's Invasion of Ukraine: February–July 2022*," Special Resources, November 30, 2022, RUSI.
4 Microsoft, "Special Report: Ukraine, An Overview of Russia's Cyberattack Activity in Ukraine," April 27, 2022.
5 Banco, Graff, Seligman, Toosi, and Ward, "Something Was Badly Wrong."
6 Zabrodskyi, Watling, Danylyuk, and Reynolds, "Preliminary Lessons."
7 Brian Whitmore, "*Did Russia Plan Its War In Georgia?*" Radio Free Europe, Radio Liberty, August 15, 2008; Kent DeBenedictis, *Russian 'Hybrid Warfare' and the Annexation of Crimea: The Modern Application of Soviet Political Warfare* (London: I. B. Tauris, 2022).
8 Banco, Graff, Seligman, Toosi, and Ward, "Something Was Badly Wrong."
9 Zabrodskyi, Watling, Danylyuk, and Reynolds, "Preliminary Lessons."
10 Ibid.
11 Dave Johnson, "*Review of Speech by General Gerasimov at the Russian Academy of Military Science*," NATO Defence College, March 30, 2019.
12 Banco, Graff, Seligman, Toosi, and Ward, "Something Was Badly Wrong."
13 Ibid.
14 Stavros Atlamazoglou, "US Special Operators Borrowed a Unique Part of Army Green Beret Training to Prepare Ukrainians to Fight Russia," *Insider*, January 24, 2023.
15 Banco, Graff, Seligman, Toosi, and Ward, "Something Was Badly Wrong."
16 Michael Mazarr, "Understanding Deterrence" (Santa Monica, CA: RAND, 2018), https://www.rand.org/pubs/perspectives/PE295.html.

17 Ibid.
18 Ibid.
19 Lawrence Freedman, "The Limits of Deterrence," in *Comprehensive Deterrence Forum: Proceedings and Commissioned Papers*, ed. Becca Wasser et al. (Santa Monica, CA: RAND, 2018), 28.
20 Justin Massie, "What Kind of Peace for Ukraine? Understanding the Causes of the War to End It," *Network for Strategic Analysis (NSA)*, November 17, 2022.
21 Jeffrey Goldberg, "The Obama Doctrine," *The Atlantic*, April 2016.
22 Massie, "What Kind of Peace for Ukraine?"
23 Ibid.
24 Banco, Graff, Seligman, Toosi, and Ward, "Something Was Badly Wrong."
25 Mazarr, "Understanding Deterrence."
26 Massie, "What Kind of Peace for Ukraine?"
27 Rob Johnson, "Dysfunctional Warfare: The Russian Invasion of Ukraine," *Parameters* 52, no. 2 (2022): 5–20.
28 Massie, "What Kind of Peace for Ukraine?"
29 Banco, Graff, Seligman, Toosi, and Ward, "Something Was Badly Wrong."
30 Andrian Prokip, "Why the Zaporizhzhia Nuclear Power Plant Matters…for the Whole World," *Wilson Center*, September 19, 2022.
31 Zabrodskyi, Watling, Danylyuk, and Reynolds, "Preliminary Lessons."
32 Vladimir Putin, "*On the Historical Unity of Russians and Ukrainians*," July 12, 2021, http://en.kremlin.ru/events/president/news/66181.
33 Verkhovna Rada Ukraine, No, 1427-XII, "Act of Declaration of Independence of Ukraine," August 24, 1991.
34 Banco, Graff, Seligman, Toosi, and Ward, "Something Was Badly Wrong."
35 Radio Free Europe, Radio Liberty, "*Latest Satellite Photos Show Increased Russian Military Activity Near Ukraine*," February 17, 2022.
36 Nikkei staff writers, "Satellite Images Unveil Russian Spearhead of Ukraine invasion," *Nikkei Asia*, February 27, 2022.
37 Sebastien Roblin, "Alleged Captured Documents Imply Ukraine Invasion Planned in January," *Forbes*, March 3, 2022.
38 Banco, Graff, Seligman, Toosi, and Ward, "Something Was Badly Wrong."
39 Volodymyr Zolkin, "Survivor: Russian VDV Airborne Soldier Talks About Their …," *Conflict Camera*, April 21, 2022.
40 Microsoft, "Special Report."
41 Brad Smith, "Digital technology and the war in Ukraine," *Microsoft*, February 28, 2022.
42 Microsoft, "Special Report."
43 Vladimir Putin, "Address by President of the Russian Federation," President of Russia, February 24, 2022.
44 Matthew Chance, "CNN reporter: This shows just how close Russian forces are to Ukraine capital," *CNN*, February 24, 2022.
45 Microsoft, "Special Report."
46 Ibid.
47 Carl von Clausewitz, *On War*, ed. and trans. Michael Howard and Peter Paret (Princeton, NJ: Princeton University Press, 1984), 370.
48 WFAA, "*Ukrainian President Volodymyr Zelensky: 'We are still here',*" February 24, 2022.
49 Putin, "Address by President."
50 Jeremy Lamoreaux, "*The Three Motivations for an Assertive Russian Grand Strategy*," in Department of Defense, Russian Strategic Intentions: A Strategic Multilayer Assessment White Paper, May 2019, 1–2.
51 Putin, "Address by President."
52 Banco, Graff, Seligman, Toosi, and Ward, "Something Was Badly Wrong.

53 Ibid.
54 Jonathan Masters, "What are Russia's Broad Interests in Ukraine? From Ukraine: Conflict at the Crossroads of Europe and Russia," *Council on Foreign Relations*, Updated February 14, 2023.
55 Jeffrey Schott, "Economic Sanctions Against Russia: How Effective? How Durable?" *Peterson Institute for International Economics*, April 2023.
56 Nick Ferris, "Weekly Data: The EU is Paying the Same Amount for 75% Less Russian Gas," *Energy Monitor*, September 20, 2022.
57 Heli Simola, "War and Sanctions: Effects on the Russian Economy," *CEPR*, December 15, 2022.
58 Andrei Kolesnikov and Denis Volkov, "The Coming Deluge: Russia's Looming Lost Decade of Unpaid Bills and Economic Stagnation," *Carnegie Moscow Center*, November 24, 2021.
59 Dan Altman, "The Evolution of Territorial Conquest After 1945 and the Limits of the Territorial Integrity Norm," *International Organization* 74, 3 (2020): 509.
60 Dan Altman, "By Fait Accompli, Not Coercion: How States Wrest Territory From Their Adversaries," *International Studies Quarterly* 61, no. 4 (2017): 886–888; Altman, "The Evolution of Territorial Conquest," 505.
61 North Atlantic Treaty Organization, "*Statement by NATO Heads of State and Government on Russia's attack on Ukraine*," Press Release, February 25, 2022.
62 Martin Wolf, *The Crisis of Democratic Capitalism* (New York: Penguin Press, 2023), 357.
63 Robert Manning, "Washington Is Missing a Chance to Turn China Against Russia," *Foreign Policy*, January 19, 2023, 5.
64 Manning, "Washington Is Missing a Chance."

Bibliography

Altman, Dan. "By Fait Accompli, Not Coercion: How States Wrest Territory from Their Adversaries," *International Studies Quarterly* 61, no. 4 (2017): 881–891.

Altman, Dan. "The Evolution of Territorial Conquest After 1945 and the Limits of the Territorial Integrity Norm," *International Organization* 74, no. 3 (2020): 490–522.

Atlamazoglou, Stavros. "US Special Operators Borrowed a Unique Part of Army Green Beret Training to Prepare Ukrainians to Fight Russia," *Insider*, January 24, 2023, https://www.businessinsider.com/us-developed-version-of-special-forces-q-course-for-ukrainians-2023-1.

Banco, Erin, Garrett Graff, Lara Seligman, Nahal Toosi, and Alexander Ward. "Something Was Badly Wrong': When Washington Realized Russia Was Actually Invading Ukraine," *Politico*, February 24, 2023, https://www.politico.com/news/magazine/2023/02/24/russia-ukraine-war-oral-history-00083757.

Bateman, Jon. "Russia's Wartime Cyber Operations in Ukraine: Military Impacts, Influences, and Implications," Carnegie Endowment for International Peace, December 2022, https://carnegieendowment.org/2022/12/16/russia-s-wartime-cyber-operations-in-ukraine-military-impacts-influences-and-implications-pub-88657.

Chance, Matthew. "CNN Reporter: This Shows Just How Close Russian Forces are to Ukraine Capital," *CNN*, February 24, 2022, https://www.youtube.com/watch?v=b40r8RX5MoE.

Clausewitz, von Carl. *On War*, ed. and trans. by Michael Howard, Peter Paret, and Bernard Brodie (Princeton, N.J.: Princeton University Press, 1984).

DeBenedictis, Kent. *Russian 'hybrid Warfare' and the Annexation of Crimea: The Modern Application of Soviet Political Warfare* (London: I.B. Tauris, 2022).

Ferris, Nick. "Weekly Data: The EU is Paying the Same Amount for 75% Less Russian Gas," *Energy Monitor*, September 20, 2022, https://www.energymonitor.ai/finance/risk-management/weekly-data-the-eu-is-paying-the-same-amount-for-75-less-russian-gas/.

Freedman, Lawrence. "The Limits of Deterrence," in *Comprehensive Deterrence Forum: Proceedings and Commissioned Papers*, ed. Becca Wasser, et al., (Santa Monica, CA: RAND, 2018), https://www.rand.org/content/dam/rand/pubs/conf_proceedings/CF300/CF345/RAND_CF345.pdf.

Goldberg, Jeffrey, "The Obama Doctrine," *The Atlantic*, April 2016, https://www.theatlantic.com/magazine/archive/2016/04/the-obama-doctrine/471525/.

Johnson, Dave. "Review of Speech by General Gerasimov at the Russian Academy of Military Science," NATO Defence College, March 30, 2019, http://www.ndc.nato.int/research/research.php?icode=585.

Johnson, Rob. "Dysfunctional Warfare: The Russian Invasion of Ukraine," *Parameters* 52, no. 2 (2022): 5–20.

Kolesnikov, Andrei, and Denis Volkov. "The Coming Deluge: Russia's Looming Lost Decade of Unpaid Bills and Economic Stagnation," Carnegie Moscow Center, November 24, 2021, https://carnegieendowment.org/2021/11/24/coming-deluge-russia-s-looming-lost-decade-of-unpaid-bills-and-economic-stagnation-pub-85852.

Lamoreaux, Jeremy W. "The Three Motivations for an Assertive Russian Grand Strategy," In: Department of Defense, Russian Strategic Intentions: A Strategic Multilayer Assessment White Paper, May 2019, https://www.politico.com/f/?id=0000016b-a5a1-d241-adff-fdf908e00001.

Liddell Hart, Basil. *Strategy* 2nd ed. (New York: Penguin Books, 1991).

Manning, Robert A. "Washington Is Missing a Chance to Turn China Against Russia," *Foreign Policy*, 19 January 2023, https://foreignpolicy.com/2023/01/19/us-china-russia-ukraine-allies-war/.

Massie, Justin. "What Kind of Peace for Ukraine? Understanding the Causes of the War to End It," Network for Strategic Analysis (NSA), November 17, 2022, https://ras-nsa.ca/what-kind-of-peace-for-ukraine-understanding-the-causes-of-the-war-to-end-it/.

Masters, Jonathan. "What are Russia's Broad Interests in Ukraine?," From "Ukraine: Conflict at the Crossroads of Europe and Russia," *Council on Foreign Relations*, Updated February 14, 2023, https://www.cfr.org/backgrounder/ukraine-conflict-crossroads-europe-and-russia.

Mazarr, Michael J. *Understanding Deterrence* (Santa Monica, CA: RAND, 2018), https://www.rand.org/pubs/perspectives/PE295.html.

Microsoft. "Special Report: Ukraine, An Overview of Russia's Cyberattack Activity in Ukraine," April 27, 2022, https://query.prod.cms.rt.microsoft.com/cms/api/am/binary/RE4Vwwd.

Nikkei Staff Writers. "Satellite Images Unveil Russian Spearhead of Ukraine Invasion," *Nikkei Asia*, February 27, 2022, https://asia.nikkei.com/Politics/Ukraine-war/Satellite-images-unveil-Russian-spearhead-of-Ukraine-invasion.

North Atlantic Treaty Organization. "Statement by NATO Heads of State and Government on Russia's attack on Ukraine," Press Release, February 25, 2022, https://www.nato.int/cps/en/natohq/official_texts_192489.htm.

Prokip, Andrian. "Why the Zaporizhzhia Nuclear Power Plant Matters…for the Whole World," *Wilson Center*, September 19, 2022, https://www.wilsoncenter.org/blog-post/why-zaporizhzhia-nuclear-power-plant-mattersfor-whole-world.

Putin, Vladimir. "On the Historical Unity of Russians and Ukrainians," July 12, 2021, http://en.kremlin.ru/events/president/news/66181.

Radio Free Europe, Radio Liberty. "Latest Satellite Photos Show Increased Russian Military Activity Near Ukraine," February 17, 2022, https://www.rferl.org/a/satellite-images-belarus-ukraine-russia-military-activity/31707879.html.

Roblin, Sebastien. "Alleged Captured Documents Imply Ukraine Invasion Planned in January," *Forbes*, March 3, 2022, https://www.forbes.com/sites/sebastienroblin/2022/03/03/alleged-captured-documents-imply-ukraine-invasion-planned-in-january/?sh=7837bb851b96.

Schott, Jeffrey J. "Economic sanctions against Russia: How effective? How durable?," Peterson Institute for International Economics, April 2023, https://www.piie.com/publications/policy-briefs/economic-sanctions-against-russia-how-effective-how-durable.
Simola, Heli. "War and Sanctions: Effects on the Russian Economy," CEPR, December 15, 2022, https://cepr.org/voxeu/columns/war-and-sanctions-effects-russian-economy.
Smith, Brad. "Digital technology and the war in Ukraine," Microsoft, February 28, 2022, https://blogs.microsoft.com/on-the-issues/2022/02/28/ukraine-russia-digital-war-cyberattacks/.
Sun Tzu. *The Art of War*, trans. by Samuel B. Griffith (Oxford University Press, 1963).
Verkhovna Rada Ukraine). No, 1427-XII, Act of Declaration of Independence of Ukraine," August 24, 1991, https://static.rada.gov.ua/site/postanova_eng/Rres_Declaration_Independence_rev12.htm.
WFAA. "Ukrainian President Volodymyr Zelensky: We are still here," February 24, 2022, https://www.youtube.com/watch?v=wgCNKhtZYks.
Whitmore, Brian. "Did Russia Plan Its War in Georgia?" Radio Free Europe, Radio Liberty, August 15, 2008, https://www.rferl.org/a/Did_Russia_Plan_Its_War_In_Georgia__/1191460.html.
Wolf, Martin. *The Crisis of Democratic Capitalism* (New York: Penguin Press, 2023).
Zabrodskyi, Mykhaylo, Jack Watling, Oleksandr V Danylyuk, and Nick Reynolds. "Preliminary Lessons in Conventional Warfighting from Russia's Invasion of Ukraine: February–July 2022," Special Resources, November 30, 2022, RUSI, https://www.rusi.org/explore-our-research/publications/special-resources/preliminary-lessons-conventional-warfighting-russias-invasion-ukraine-february-july-2022.
Zolkin, Volodymyr. "Survivor: Russian VDV Airborne Soldier Talks About Their …," Conflict Camera, April 21, 2022, https://www.youtube.com/watch?v=TStvtOgp4ow&t=177s.

4 Operation Orchard

The Israeli Strike at *Al-Kibar*

Introduction

The second case study in this book describes a preventive strike that was executed by the Israeli Defense Force (IDF). In 1981, during Operation Opera, the IDF attacked an unfinished Iraqi nuclear reactor. Although the worldwide reactions were highly critical, the operation resulted in the "Begin doctrine," which became a synonym for an Israelian surgical strike to prevent their adversaries from acquiring weapons of mass destruction (WMD). Although the Israeli strike successfully destroyed the Osirak nuclear reactor, it did not stop Iraqi proliferation but somewhat accelerated the program and steered it in a covert direction. In other words, although the Israelis achieved the operation's desired tactical objective, it's doubtful whether the political objective of severely disrupting the Iraqi nuclear program was achieved as well.

This chapter focuses on Operation Orchard, a comparable operation executed nearly three decades later, paying particular attention to the potential of modern warfare. From a strategic perspective, the various components required the operation to function as an integrated system. From a tactical perspective, the operation included the synergistic employment of capabilities across the air, land, and space domains, plus the electromagnetic spectrum (EMS), all of which played a vital role in the operation's outcome.

Background

When former Syrian President Hafez al-Assad died in office on June 10, 2000, his son Bashar succeeded him on July 17, 2000. Bashar was then only 34 years old and had little political experience. After completing his medical studies at Damascus University in 1988, Bashar started a career as a physician in the Syrian Armed Forces. By 1992, Bashar moved to the Western Eye Hospital in London, where he specialized in ophthalmology. US intelligence paid close attention to the new president and, contrary to his father, considered him a weak leader.[1] Indeed, the Israelis had considered Hafez al-Assad to be a "political realist," and shortly before his death was close to reaching an agreement with him over the Golan Heights.[2] Bashar, by contrast, was assessed by the Israelis as unpredictable, and his regime

DOI: 10.4324/9781003365112-6

was believed to pose a significant threat to Israel.[3] The new al-Assad regime, for example, reportedly stepped up its support to Hezbollah, Iran's Shi'a militia proxy force that fought the Israelis in Lebanon (see Chapter 7).[4] The regime also invited high-ranking delegations from Pyongyang, North Korea, to Damascus, with whom it closely collaborated in developing the capabilities of the Syrian Armed Forces, including medium-range ballistic missiles and chemical weapons.[5] Bashar al-Assad allegedly went even further than his father by starting his own program to develop nuclear weapons.[6] In retrospect, however, the US Central Intelligence Agency (CIA) had assessed that the decision to start this program was made in the late 1990s. A decision thus made by Hafez al-Assad.[7]

The first signs of this nuclear threat became visible in 2001 when US intelligence collected material about a potential nuclear partnership between Syria and North Korea.[8] The material included reports on senior North Korean officials associated with the Yongbyon nuclear reactor who frequently visited Syria and North Korean representatives obtaining materials for an unknown facility in Syria.[9] From the Syrian side, General Muhammad Suleiman, commander of Syria's clandestine operations, coordinated the cooperation between the two countries. Suleiman, who also was responsible for cooperation with Iran and Hezbollah, reported directly to President Bashar al-Assad.[10]

To better understand the stage of the Syrian nuclear program, US intelligence agencies paid specific attention to a remote, unidentified, large cubical building in the eastern *Deir ez-Zor* region of Syria, referred to as *Al-Kibar*. According to former CIA Director General Michael Hayden, the US National Geospatial-Intelligence Agency (NGA) reported on the site for the first time in 2004. Still, the NGA could not identify its purpose at that time.[11] During the same year, the US National Security Agency (NSA) collected various encrypted electronic messages exchanged between Syria and North Korea but could not decipher them.[12] Given the significance of the intelligence, the Americans shared their findings with their Israeli counterparts.

Meir Dagan, head of Mossad, was skeptical about the American findings. Also, Amnon Sufrin, director of the intelligence division at Mossad, displayed similar reservations by explaining that Israeli intelligence's main focus was Iran, and most analysts assessed the possibility of a Syrian nuclear program as unlikely. However, Sufrin decided to assign a team to study the American findings because he believed that if the Syrians were hiding something, the team would find it.[13] Members of the research branch of Military Intelligence joined the team. They could not logically explain the unidentified site in the middle of a remote Syrian desert. Therefore, the Israelis prioritized the site for intelligence collection.[14]

Despite the increased efforts to collect intelligence on the site, it took approximately two years until the Israelis achieved their first breakthrough. In a document known as "An Issue for Examination," the analysts presented an alternative explanation for the site's real purpose.[15] Rather than supporting the accepted assumption that the site enriched uranium using centrifuges, the team suggested that the installations were meant to produce plutonium.[16] In January 2007, the Israelis achieved a second breakthrough when Imagery Intelligence (IMINT) showed a pipeline leading from the building toward the Euphrates. Such a pipeline indicated the presence of a cooling system critical for producing plutonium.[17] Moreover, the existence of a cooling system made Israeli intelligence inclined to believe the findings as presented in the document "An Issue for Examination."[18] Nevertheless, the Israelis were still not confident enough to strike the site but rather wanted to collect additional material to reveal the site's true purpose. To this end, the Israelis conducted several operations abroad.[19]

A Mossad operation executed in March 2007 in Vienna stands out. During the operation, the Israelis collected a significant quantity of material about the site, including blueprints and a substantial number of photos of the installations.[20] In one photo, Ibrahim Othman, the head of the Atomic Energy Commission of Syria, and Chon Chibu, the head of North Korea's Yongbyon nuclear reactor, were seen together at the site.[21] US intelligence officials were the first who openly referred to the material obtained by Mossad in Vienna in a "Background Briefing" with Senior US officials held on April 24, 2008.[22] Sufrin also referred to this material when he recalls that the head of Mossad informed Prime Minister Ehud Olmert that the site was a plutonium-producing nuclear reactor.[23] Finally, Olmert accepted the authenticity of the material and found it to be a game changer as he recalled, "Meir [Dagan] came to me with this material (the photos taken from Othman's laptop) .... and it was like an earthquake. I realized that from now on everything would be different."[24]

In mid-April, Dagan traveled to the US and met with US Vice President Dick Cheney, National Security Advisor Stephen Hadley, and his deputy Elliott Abrams. During the meeting, Dagan presented the photographs of the installation and the photos of Chibu. Even though US intelligence had detected that Chibu frequently visited Syria, these photos were the first concrete evidence that North Korea was proliferating its nuclear capabilities in the Middle East and assisting Syria in building a nuclear reactor. The Americans were shocked, and Cheney immediately remembered President Bush's viewpoint on North Korea's nuclear capabilities. The President had said: "The transfer of nuclear weapons or material by North Korea to states or non-state entities would be considered a grave threat to the United States, and we would hold North Korea fully accountable for the consequences of such action."[25] When reflecting on Bush's statement, Cheney wondered why North Korea was more involved with transferring material than constructing the nuclear reactor. Dagan concluded his presentation: "This has got to go away" and added that "Syria cannot be allowed to have nuclear weapons."[26]

The next day, Dagan met with Hayden at the CIA Headquarters to discuss the installations at *Al-Kibar*. Hayden was surprised by the significance of the material. He also asked himself what the reason was for Dagan's generosity and concluded that he tried to use him to influence US foreign policy decision-making regarding the case.[27] In other words, Hayden was afraid that Dagan wanted to persuade him to use the material to get buy-in from the Bush administration, which, in turn, might have led to a direct US intervention.

Hayden felt he had to act upon the material shown to him and quickly organized a meeting, including the participation of some of his most senior intelligence officers. During this meeting, the CIA officials discussed the question of whether the material was authentic or perhaps highly sophisticated forgeries.[28] Whereas the CIA assessed the material almost exclusively as authentic, some of the agency's officials expressed reservations regarding one specific photograph that showed a written text on the side of a truck that could be forgery. Hayden did not think the Israelis would have made such a forgery themselves. Still, because Dagan did not share his source of the material, the American reservations made Hayden decide to investigate the authenticity of the material and ask a team of analysts to determine whether the material was genuine.[29] Shortly after that, a three-dimensional computerized model of the installations at *Al-Kibar* was built and compared to IMINT data collected from other sources. The result was shocking as it was a perfect match.

The next day, Hayden attended a meeting at the White House where he presented the Israeli material and the CIA's findings to President Bush. Despite the significance of the material, President Bush was hesitant to intervene, and he ended the meeting by saying to the CIA director: "Gotta be secret, and gotta be sure."[30]

Hayden followed up on the President's advice by quickly bringing together a team to investigate the validity of the material. According to Condoleezza Rice, then US Secretary of State, the team's mission was "an intensive, months-long effort to confirm and corroborate the information Israel provided us on the reactor and to gather more details from our own sources and methods."[31] For example, the team compared the Israeli photo material of the installations at *Al-Kibar* with some of their own material obtained from the Yongbyon plutonium-producing reactor in North Korea. They assessed that both installations were similar in size and capacity.[32] Interestingly, according to the CIA, North Korea was the only country that had built such nuclear installations in the last 35 years.[33] After months of research, it was concluded that the installations at *Al-Kibar* were the same as the North Korean Yongbyon installations.[34]

Despite the high degree of certainty with which the team presented its conclusions, Hayden wanted to make an additional effort to challenge these findings; Hayden, therefore, formed a so-called "Red Team." The Red Team's members had not yet been previously involved in the case and thus were unfamiliar with the material. Hayden asked the team to "build an alternative case as to why it's not a nuclear reactor; why it's not intended to produce plutonium for a weapon; why North Korea is not involved."[35] After a detailed analysis of all materials available, the Red Team concluded that "if it isn't a nuclear reactor, it must be a fake nuclear reactor."[36] In other words, if it's a hoax, it must be an act of deception. Given that the Red Team found the hoax scenario highly unlikely, they confirmed to Hayden that the installations at *Al-Kibar* were indeed a secret Syrian nuclear reactor to produce plutonium. Hayden subsequently informed President Bush about their findings but added a reservation since the CIA could not identify other critical components of a nuclear weapons program, such as developing a nuclear warhead. This reservation was of great significance, as Syria is a signatory of the Nuclear Non-Proliferation Treaty (NPT), which does not allow them to develop nuclear weapons. Consequently, the level of analytic confidence applicable to the CIA's findings was incomplete.[37]

Shortly after Hayden presented the case to Bush, a small group of senior US Cabinet members proposed three distinct response options to him. First, Vice President Cheney proposed a kinetic strike that would destroy the installations and send a strong message to rogue regimes that may operate or start similar clandestine nuclear programs. Rice, by contrast, did not favor the military option but rather favored diplomacy to force the Syrian regime to dismantle the installation. Finally, Robert Gates, who then served as the Secretary of Defense, agreed that a kinetic strike would be appropriate but suggested that Israel should conduct the strike itself. [38] On June 19, 2007, President Bush made his decision and told Olmert during a meeting in the White House that the US could not support Israel by executing the strike. Bush explained to Olmert that his decision was mainly based on a lack of evidence. The fact that Bush gave this explanation is noteworthy, given that the Israelis recently had updated the already significant intelligence available to him with additional material collected by *Sayeret Matkal*, an Israeli Special Operations Forces (SOF) unit, which had taken photographs of the site and collected soil samples to confirm that the site was indeed a nuclear one.[39]

After the meeting, Olmert recognized that if he wanted to destroy the installations at *Al-Kibar*, Israel would have to plan, prepare, and execute a mission all by itself. Immediately upon return to Israel, Olmert directed the IDF to formulate options to destroy the installations. The IDF presented three plans: air strikes against several sites connected to the Syrian nuclear arms program, an isolated surgical air strike targeting the installations at *Al-Kibar*, and a SOF ground raid.[40] The latter option was planned and prepared in great detail by *Sayeret Matkal* and included fully dressed rehearsals on a replica of the installations constructed in an undisclosed location in the south of Israel. The Israelis considered strikes against multiple targets to be complex and involved a high risk of collateral damage. The use of Israeli SOF would also include significant political- and tactical risk. Moreover, the photo material taken from Othman's laptop and soil samples collected by *Sayeret Matkal* did not require the Israelis to conduct an on-site Technical Exploitation Operation (TSO). These considerations made Olmert decide to select a precision strike planned, prepared, and executed by the Israeli Air Force (IAF) to destroy the installations at *Al-Kibar*.[41]

The nuclear installations at *Al-Kibar*

The material obtained by the Israelis indicated that the Syrians began with the construction of the nuclear installation as early as 2000.[42] The Syrians received significant support from North Korea, which operated a similar installation typically referred to as the Yongbyon installation at that time. Based on the material available, US intelligence analysts had testified in a 2008 Congressional hearing that the Syrians had completed the construction in the summer of 2007.[43]

The installations, depicted in Figure 4.1, were situated in the remote eastern *Deir ez-Zor* region of Syria, an area referred to as *Al-Kibar*. The installations comprised a main square building of approximately 1,600 square meters and 20 meters high, which lay a kilometer east of the Euphrates. The Syrians, clearly aware of

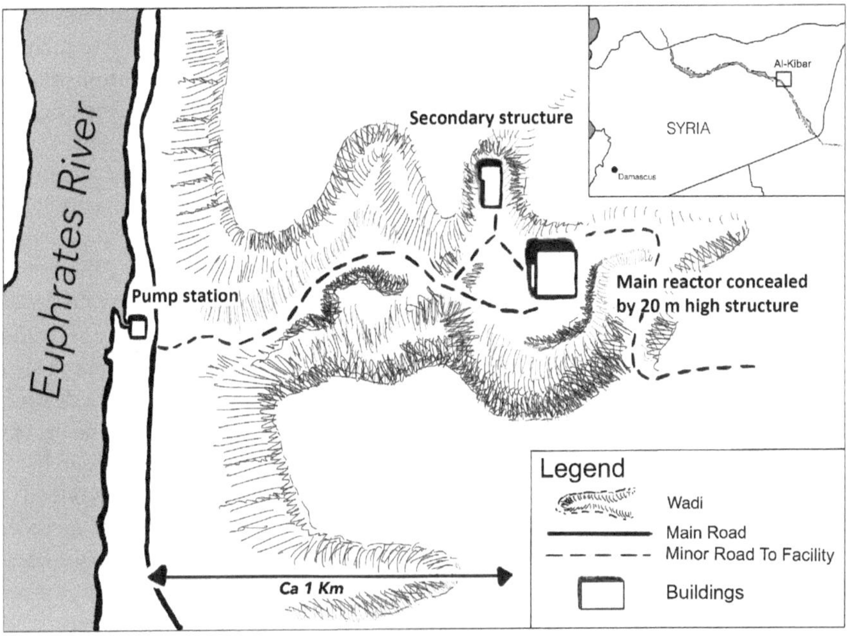

Figure 4.1 Overview of the nuclear installations at *Al-Kibar*.

Western aerial and orbital reconnaissance capabilities, tried to conceal the site by building the installations in a *wadi* (i.e., a dry riverbed) and covering them with a camouflage structure that changed the site's profile and prevented anyone from capturing its design from above. The camouflage structure, however, made the Israelis suspicious, as there would be no plausible explanation for the presence of such a building in a deserted area where only a few herdsmen were living.[44]

The area around the main building was scattered with rubble. It wasn't equipped with physical security measures such as fences, guards, or ground-to-air defenses to protect the installations from sabotage or air strikes. Intelligence had assessed that the absence of any security measure was intentional: the Syrians wanted to deceive potential onlookers and hide the site's true nature.[45] CIA analysts concluded that "when the pipeline and pump house were externally completed" at the beginning of August 2007, the facility would have been nearing operational capability to start producing plutonium, the critical component of producing nuclear weapons.[46] The Israelis, therefore, had to move fast.

Planning and preparation

The planning and preparation for the operation began with a detailed analysis of the possible courses of action for destroying the installations. Apart from the destruction of the installations, the Israelis wanted to provide al-Assad with the possibility to deny the presence of a nuclear installation on Syrian territory rather than embarrass and humiliate him publicly. Such an approach, the Israelis assessed, would also prevent al-Assad from responding aggressively by launching a counterattack on Israeli soil. According to Israeli intelligence, this criterion implied that the strike had to be discreet and executed with a small footprint. Such discretion would allow Israel to deny its involvement and Syria to downplay the event.[47]

During the planning of the operation, the element of surprise was given top priority, while the size of the strike force was up for debate. By mid-June 2007, the IDF started planning the operation. Drawing from the scenario presented by Israeli intelligence, the principle of surprise had to be the crucial element of the operation.[48] To fully exploit the element of surprise, two different options were developed wherein the size of the strike force served as the distinctive feature. The first option included a strike force that had a vast number of warplanes. Although such a large force package would provide a high level of certainty regarding the destruction of the site, it also might provoke an aggressive Syrian response. The second option, by contrast, included a force package of just eight warplanes–four F-15I "*Raam*" (Thunder) and four F-16I "*Sufa*" (Storm) carrying a combined payload of 20 tons of bombs. The central idea behind this small force package that included two different strike aircraft was to create redundancy while simultaneously keeping a small footprint. After comparing both options, the Israelis decided to move forward with the second option.[49]

Given its kinetic character, the Israelis determined that it would be acceptable for the operation itself, or at least its main effect (i.e., the destruction of the site), to be detectable. By contrast, the significant political and military-strategic interests and risks involved in the operation made the Israelis decide that their role as sponsors had to remain secret. Thus, the operation had a covert character, reaching above the detection threshold but staying below the attribution threshold. To this end, the planning, preparation, infiltration, and exfiltration phases of the operation were designed to avoid detection. The kinetic strike against the target, however, would be detectable. Importantly, the covert design of the operation would prevent the Syrians or other actors from tracing back the action against the target to the sponsor, Israel.

To this end, the Israelis applied various measures to ensure operational security (OPSEC). For example, the Israeli agencies involved in planning the operation used different code names for it: the IAF referred to "Soft Melody," the Defense Staff used the code name "Ein Habasor," and MI referred to Operation "Outside the Box." Besides the involvement of these agencies, the IAF HQ only informed the fighter squadrons' leadership, which was tasked with executing the strike. The IAF also did this in a limited manner, trying to reveal as few details of the target as possible. According to Colonel Amir, one of the fighter pilots who flew the mission, this high level of secrecy caused significant challenges to the operational squadrons as they had to prepare for the upcoming operation and the potential hostilities that might break out as a result.[50] The preparations included a variety of Tactics, Techniques, and Procedures (TTPs) that caused the crews to become more aware of the significance of the upcoming operation.[51]

As time progressed, Amir explained that more detailed information about the operational planning, including the Operational Environment (OE) and details of the target, became available to the crews of the fighter squadrons.[52] Parts of the information were of grave concern to the crews as various complications could hinder the execution phase, including the possibility of air defenses that had remained undetected by Israeli intelligence and the chance that a single sortie would not be sufficient to destroy all installations effectively. The latter would mean that a second sortie would be required to generate the desired effect. Making a second sortie would have significant implications for the fighter's freedom of movement, as the first sortie would likely compromise the operation and alert the Syrian Defense Force, including its fighter capacity.

Amir recalls that the crews were able to address all contingencies present in their tactical plan through the application of the operational factors of time, space, and force:

> We had a number of kinds of weapons, a mix that wouldn't leave you with a single malfunction point without the ability to compensate for it. If a certain weapon or one of the platforms didn't work, there would be something to cover for it .... In manning the crews, we choose the best. The squadron commander and his deputy take the ones on whom they agree hands down will get the bombs to the target.[53]

The crews' plan ensured the requisite redundancy, while the crews themselves consisted of the best personnel within the IAF.

Given the risk of escalation, the IDF also had to prepare some other components and increase their readiness. To maintain OPSEC, however, these components were provided with a cover story that would also be used to inform the Israeli public and the international media. According to Amir, the narrative consisted of a scenario, including a confrontation between Syria and Israel that could break out due to the recent hostilities in Lebanon:

> The explanation that was offered was concern about the possibility of hostilities breaking out over mutual miscalculations: Israel and Syria are suspicious of each other and exchanging threats in the context of the results of the last war in Lebanon. The tension is liable to lead to an unplanned deterioration in the situation, and in order to prepare for that possibility the IDF must increase its level of readiness. [54]

By the end of August 2007, the planning and preparation phases of the operation were completed, and the fighter squadrons were ready to conduct the unknown operation they had been planning and preparing over the last couple of months.

The strike

On the evening of September 5, 2007, the force package assembled on the Ramon and Hatzerim Air Bases in southern Israel, where the IAF made the final preparations for the strike. Up to this point, most of the personnel involved in the planning and preparation for the operation had not been told its objective. At 10.30 p.m., the IAF strike force, consisting of the four F15 and four F16 fighters, took off from both airbases and set out for their target.[55] They would first fly northwest along the Mediterranean and then follow the Turkish-Syrian border before making their final leg eastward toward the target. Amir recalls it was a very dark night, and the fighter aircraft flew approximately 200 feet (i.e., 66 meters) above ground level (AGL). Flying at such a low altitude required the crews to make instant decisions as various unpredictable factors could impact their clandestine approach. To counter any unforeseen events, the crews had planned for multiple contingencies. As a result, they were not required to make any transmissions while in flight, thereby reducing the risk of being intercepted by Syrian SIGINT.[56]

The Syrian air defenses were another major threat to the crews while flying in hostile territory. According to Israeli military analysts, the IDF's employment of virtual capabilities to compromise the availability of the Syrian integrated air defense system (IADS), centered around Russian equipment such as the Buk-M1 SAM missile system, played a decisive role in the operation.[57] Similar findings indicate that the IAF successfully disrupted a Syrian radar site near *Tall Al Abuad,* situated close to Turkey's Southern border, by employing advanced electronic warfare (EW) capabilities.[58] The fact that Syria's IADS, generally considered one of the more sophisticated systems available at that time, did not detect the Israeli strike force from entering and leaving Syrian airspace, while at the same time, the Syrian electrical grid remained untouched, indeed indicates the successful use of advanced cyber and EW capabilities.

Although many of the details around the employment of these advanced capabilities remain highly classified, several authors suggest that the Israelis employed them in the form of an electronic attack (EA).[59] Israeli air-to-ground EW capabilities are similar to the Suter system, a US military capability that performs a network invasion function.[60] At the time of the strike, the third generation of Suter was available to a limited number of countries. The first generation provides the capability of intercepting enemy radar imagery. Suter's second generation goes further, allowing one to take over the radar system's sensors. The third generation can gain complete control of the system, including rocket launchers and radars. The Israelis have reportedly modified two Gulfstream G550 business jet aircraft into special operations air assets to employ this sophisticated capability.[61] These findings are supported by observations that in addition to the eight-fighter aircraft, several other aircraft also took part in the operation. These aircraft headed in a northwest direction, with the task of manipulating Syrian IADS.[62] Other authors suggest that the fighter aircraft themselves were equipped with the third generation Suter software.[63]

A second sophisticated attack capability might have been using a "kill switch" that was deliberately built into the IADS to paralyze its function.[64] A chip can be manipulated in two distinct ways: exploiting a back door or activating a kill switch.[65] A kill switch is a mechanism to shut down or disable a specific piece of software or hardware, while a back door refers to gaining high level access to software or hardware by bypassing routine access control measures.[66] Although manipulating chips is expensive and requires significant technical skill, it is technically possible.[67] At the same time, a chip's complexity and industry inspection regulations make it very difficult to determine whether a chip has been modified.[68] By applying advanced planning and making the required investments, Israel may have been capable of manipulating chips that were part of the Syrian IADS.[69] Whether the Syrian IADS was manipulated using EW or cyber-attack, the execution of the operation indicates that the IDF could effectively disrupt the Syrian air defense systems and enter its air space undetected. In doing so, the Israelis were able to minimize collateral damage, as a kinetic strike on Syrian radar sites would have likely resulted in the death of Syrian soldiers and perhaps even civilians.[70]

The Israelis maintained radio silence while infiltrating Syrian airspace to ensure avoiding detection. Faced with only minor technical issues that the individual crew members could solve, the strike force could start its final approach to the installations at *Al-Kibar* undetected. Given that the fighter aircraft were still flying below 200 feet AGL, they now had to climb to higher altitudes that would allow them to employ their weapon systems. Amir recalls:

> And then you get to the target area, climb a bit higher, pulling into execution mode. Everything goes according to the advance briefing. Each plane drops two bombs. You feel a slight shaking of the wing as they are released. There is a bit of excitement before the bomb hits. We made certain five times that it was 'all systems go.'[71]

From his cockpit view, Amir remembers the image when the bombs were dropped on the installations. The strike force remained over its target only for two to three minutes, during which each fighter aircraft reported two direct hits on the target. On September 6, at 00:25 a.m., after the last bomb had been released and its impact confirmed, the flight lead sent the codeword "Arizona" back to the Israeli operations center, indicating that the target had been destroyed.[72]

Immediately after that, the strike force started its exfiltration, which was still full of risk, as emphasized by Amir: "We had hundreds of miles to fly, some of them in an area full of missiles. We were very focused on the escape route because it could elicit surprises."[73] Indeed, the mission up to this point proceeded as planned. Yet, during the exfiltration phase, an unexpected event occurred, as one of the fighter aircraft had to save fuel for the long exfiltration and decided to release one of the plane's external fuel tanks. The fuel tank, unfortunately, landed on Turkish soil. From a tactical perspective, releasing the fuel tank was a justified decision. From a strategic perspective, however, the IAF considered the decision a significant operational misjudgment. What followed was an incident as photos of the fuel tank revealed that it had Hebrew writing on it.[74] Hence, the mission's covert character was compromised since the evidence indicated Israeli warplanes had been flying in the area during the time of the strike. Moreover, since the Israelis had not informed the Turkish government about the operation, Israel quickly sent Amos Yadlin, Director of Military Intelligence, to Turkey with the task of explaining the circumstances and assuring them that Turkish sovereignty was not violated during the operation.

Apart from this incident, the exfiltration was conducted without further difficulty, and all aircraft safely returned to Ramon and Hatzerim Air Bases at 2:30 a.m.

The next day, Syrian officials reported that its Air Defense System had registered an intrusion of Syrian Air Space by Israeli warplanes. According to a Syrian military spokesperson, "the Israeli enemy aircraft infiltrated into the Arab Syrian territory through the northern border, coming from the Mediterranean, heading toward the eastern region, breaking the sound barrier."[75] The spokesperson did not mention anything about the strike at *Al-Kibar* but claimed that "air defense units confronted them and forced them to leave after they dropped some ammunition in deserted areas without causing any human or material damage."[76] During the following days, the Syrian government claimed to consider an appropriate response to the Israeli strike.[77] The fact that such a response did not occur confirmed the Israeli assessment that the al-Assad regime would be reluctant to do so because of fear that the nature of the installations at *Al-Kibar* would be exposed.

The Israelis also wanted to keep the operation secret and did not release anything to the press. Olmert, for example, stated, "I don't know what you are talking about" when asked by *Haaretz* to respond to the Syrian claims.[78] The Israelis employed, however, intensive diplomatic efforts to inform friendly nations in the region and other Western countries about the nature of the operation and gain support for the strike. In a private phone call, Olmert also informed President Bush about the operation: "Oh, that is very interesting" was Bush's response to the news.[79] Five days after the strike, US Defense officials confirmed that Israeli warplanes had conducted an air strike in northeastern Syria. The exact nature of the strike remained unclear to them, however.[80] Finally, Bashar Jaafari, the Syrian ambassador to the United Nations (UN), informed the UN Security Council (UNSC) about the Israeli violation of Syrian airspace but noted in his letter to its members that the dropped munitions did not achieve any physical damage. However, when the UNSC discussed the event, its representatives could not draw any conclusions. Although reports about the event were then circulating in the media and discussion was taking place within diplomatic spheres, the exact circumstances of what had happened in Syria on the night of September 6, 2007, would remain uncertain for years to come.

Post-strike activities

Shortly after the Israeli strike on the installations at *Al-Kibar*, Syrian authorities completed the destruction of the facility to cover up the affair. Indeed, as several media reports about the site and the strike appeared over time, the Syrian authorities claimed that the *Al-Kibar* area is simply a remote piece of desert with no significant infrastructure whatsoever. However, when International Atomic Energy Agency (IAEA) inspectors visited the site in 2011 during the Syrian Civil War, they took samples that indicated traces of radioactive material. As a result, the IAEA concluded that it was highly likely that a nuclear facility had been present in the area. When confronted with the IAEA report, the al-Assad regime denied it. Israel also denied its involvement in the operation and continued to do so for more than a decade.

The operation only became official on March 14, 2018, when Israeli authorities released declassified material about the operation.[81] Simultaneous with the release of these materials, former Israeli Prime Minister Benjamin Netanyahu announced the operation on Twitter: "The Israeli government, the Israel Defense Forces and the Mossad prevented Syria from developing nuclear capability. They are worthy of full praise for this. Israel's policy was and remains consistent - to prevent our enemies from arming themselves with nuclear weapons."[82] When he exposed the operation, Netanyahu made no public statements regarding North Korea's involvement, nor did its participation, which was significant, come, at least overtly, at a cost for the regime in Pyongyang. For Amir, Netanyahu's statement and the release of the material meant that his wife would finally know what he had kept secret from her for more than a decade: "Efrat will know about it only now, when the articles and interviews are published. There are things about which you do not talk. She is familiar with this."[83]

Analysis

Policy

Given the magnitude of Syria's nuclear threat to Israel, destroying the site at *Al-Kibar* was vital to protecting Israel's national interests. For decades, the al-Assad regime had seriously threatened Israel and the wider region. Syria's support to Hezbollah, Iran's Shi'a militia proxy force, for example, is a destabilizing factor in Lebanon, a country that directly borders Israel. Furthermore, Syria's cooperation with the regime in North Korea enables both countries to develop their military capabilities, including medium-range ballistic missiles and chemical weapons. With the development of a nuclear weapons program, the al-Assad regime did not only violate the NPT but also posed an immediate threat to Israel. When the Syrians completed the construction of the installations at *Al-Kibar* in the summer of 2007, US intelligence assessed it would reach operational capability by August 2007.[84] Israeli intelligence estimated that from that moment on, the Syrians would have been able to start with the production of plutonium, a critical component in a nuclear arms program. According to an Israeli intelligence report:

> Syria has set up, within its territory, a nuclear reactor for the production of plutonium, through North Korea, which according to an (initial) worst-case assessment is liable to be activated in approximately another year. To our assessment [REDACTED] secretive and orderly [REDACTED] for achieving a nuclear weapon.[85]

The Israelis believed the imminent threat posed by the reactor had to be eliminated. Olmert gave its military the order to develop various options to destroy the site to achieve that strategic objective.

Military strategy

The Israelis chose a precision strike because they prioritized minimizing the risk of collateral damage and maintaining a low profile to prevent Syrian retaliation. To achieve the strategic objective of destroying the installations at *Al-Kibar*, Olmert asked the IDF to develop various options. They presented him with three plans: air strikes against several sites connected to the Syrian nuclear arms program, an isolated surgical air strike targeting the installations at *Al-Kibar*, and a SOF ground raid. During the interagency decision-making process, the Israelis prioritized minimizing the risk of collateral damage and maintaining a low profile. As a result, Olmert quickly dropped the option of a strike against multiple targets. The Israeli cabinet assessed the risks of a SOF raid and considered them too high. Olmert, therefore, decided that the installations at *Al-Kibar* had to be destroyed by a precision strike, planned, prepared, and executed by the IAF. Whether the strategy was inherently logical depends on its suitability, feasibility, and acceptability.

Suitability: could the Israelis execute the operation with the available resources?

Entering well-defended enemy airspace with a strike force powerful enough to destroy the nuclear installations and safely returning to base while remaining undetected is highly complex and challenging. The Israelis, however, could do so by having superior intelligence and employing a combination of advanced EW capabilities and superior TTPs.

The Israelis conducted a significant shaping phase before the strike that included a variety of intelligence collection capabilities (e.g., Computer Network Operations (CNO), HUMINT, IMINT), and SIGINT). The collected intelligence allowed the Israelis to assess with a very high level of certainty that the installations at *Al-Kibar* were indeed a Syrian nuclear reactor. This finding gave the Israelis the legitimacy to conduct the strike and follow up with diplomatic efforts to satisfy their partners and allies regarding the reliability of their intelligence. The specific intelligence about the Syrian IADS allowed the Israelis to develop the required EW capability to manipulate this system effectively.

The IAF knew that infiltrating Syrian air space undetected, even while flying at a very low altitude, would be a major tactical challenge. During the planning phase, the IAF concluded that next to the manipulation of Syrian IADS, the strike force had to be small to reduce its detectability by radar. Furthermore, the strike force had to incorporate redundancy that guaranteed the Israelis a high level of certainty that a minimum number of fighter aircraft would reach and destroy the installations.

Feasibility: did achieving the objectives create the strategic effects contributing to the desired end?

The strategic objective for the operation was to destroy the nuclear installations at *Al-Kibar*, which were nearing operational capability at the beginning of August 2007.

By destroying the installations, the Israelis created a strategic effect that contributed directly to the desired end, as the loss of the installations led to a significant setback in the Syrian nuclear program. Hence, the nuclear threat would become less imminent. However, the destruction of the site was a solution in the near term and would allow Israel, with its allies and partners, to develop a long-term solution to the Syrian nuclear threat. In other words, by conducting the operation, the Israelis bought time. The operation's success also contributed to an additional strategic objective, namely maintaining a favorable balance of power in the region, as the operation demonstrated Israel's ability to successfully employ military force surgically and with complete surprise.

Acceptability: did the strategic effects sought justify the objectives pursued and consider tangible and intangible costs?

The strategy was acceptable because the operation's discretion and strike force's small footprint allowed al-Assad to deny the presence of the nuclear installations and avoid loss of face, thereby mitigating Syria's need to escalate with a counterattack. In case of escalation, however, the Israelis had prepared some of its other IDF components and increased their readiness.

The strategy was also acceptable because it allowed Israel to maintain its favorable balance of power in the region. To this end, Israel had to conduct the operation covertly. Special operations are conducted covertly when major politico-strategic and military-strategic interests and risks are involved, requiring the relevant sponsor to remain secret. A covert character was most suitable given the purpose, target, and attack method selected for the operation. Therefore, the planning, preparation, infiltration, and exfiltration phases of the operation were designed to avoid detection. By contrast, the actions on the objective (i.e., the destruction of the installations at *Al-Kibar*) were designed to make the operation's main effect detectable. Inherent in the character of a covert operation, the IDF applied measures of OPSEC that prevented the Syrians from tracing back the action to Israel (i.e., the relevant sponsor). Although releasing the external fuel tank compromised Israel's involvement, the operation's discrete footprint did not persuade the Syrians to retaliate.

In conclusion, the inherent logic of the strategy developed by Israel was valid. The ends, ways, and means answer the questions of suitability, feasibility, and acceptability. The Israelis employed strict measures of OPSEC throughout the entire operation. By doing so, the Israelis successfully concealed the planning and preparation phases. Furthermore, using advanced EW capabilities and superior TTPs allowed the strike force to enter Syrian airspace undetected, destroy the site, and safely return to base. Despite the measures applied, however, the decision made by one of the crews to release an external fuel tank exposed Israel's role in the operation. Although Israel's role was revealed, it was able to achieve its strategic objective and thereby advance its national interest of reducing the nuclear threat posed by Syria.

Risk

What follows is an assessment of the risks associated with the operation.

Which assumptions did the Israelis make in their strategy, and what was the effect of any wrong assumption?

The Israelis correctly assumed that the employment of their advanced EW capabilities would allow them to operate in Syrian airspace undetected. A significant threat to the Israeli strike force flying into hostile territory was the Syrian IADS, centered around Russian equipment such as the Buk-M1 SAM missile system. The IADS could immediately detect the Israeli warplanes entering Syrian airspace, thereby compromising the operation at an early stage. Furthermore, detection might have led Syria to deploy its fighter aircraft for air interdiction. A mid-air confrontation between the Syrian and Israeli fighters would have likely resulted in a failed mission, as such a confrontation consumes a large amount of fuel, and the Israeli aircraft were not prepared and equipped for such a mission. More dangerously, Syria could have decided not to deploy its air interdiction capability but rather engage the Israeli strike force with the BUK-M1 SAM missile system. This scenario also would have likely resulted in mission failure. The Israelis, therefore, successfully employed advanced EW capabilities that allowed them to gain complete control of the Syrian IADS, including its radars. Controlling the IADS allowed the Israelis to manipulate the system, which, in turn, enabled the Israeli warplanes to remain undetected for the entire duration of the execution phase of the operation.

Did the Israelis incorporate flexibility or adaptability into their strategy? If required, how could they have modified their strategy, and at what cost?

The essence of the plan to destroy the installations at *Al-Kibar* did not change after the IDF had presented it as one of three possible courses of action. Out of fear of compromising the operation, the Israelis kept it secret from most personnel involved and did not conduct a fully dressed rehearsal. Instead, the Israeli air crews extensively practiced TTPs between June and September 2007. Despite the lack of a fully dressed rehearsal, the operation was conducted according to plan, with only one notable exception: releasing the external fuel tank.

How did other actors react to the Israeli operation?

Because the precision strike's strategy had resulted in very limited perceptibility, the operation received little attention from the media at the time. Some newspapers picked up on the Syrian reports on Israeli warplanes that had violated Syrian airspace. When *Haaretz* asked Olmert to respond to the Syrian claims, he answered, "I don't know what you are talking about."[86] At the same time, the Israelis employed intensive diplomatic efforts to inform friendly nations in the region and other Western countries about the nature of the operation to gain support for the strike and prevent any undesirable responses from the international community.

Finally, Bashar Jaafari, the Syrian ambassador to the UN, informed the UNSC about the Israeli violation of Syrian airspace but noted in his letter to its members that the dropped munitions had not resulted in any physical damage. When the event was discussed in the UNSC, its representatives could not draw any conclusion, and no formal response was given.

How did unforeseen factors play a role in the strategy?

Most of the operation proceeded as planned, yet an unexpected event occurred during the exfiltration phase. One of the fighter aircraft had to save fuel for the long exfiltration and decided to release one of the plane's empty external fuel tanks. The fuel tank, unfortunately, landed on Turkish soil. From a tactical perspective, the release of the fuel tank was fully justified. From a strategic perspective, however, the IAF considered the decision a significant operational misjudgment. A diplomatic incident followed as the fuel tank had Hebrew writing on it. Hence, the mission's covert character was compromised since the material indicated Israeli warplanes had been flying in the area during the time of the strike.

Tactics

From a tactical perspective, this is one of the most successful precision strike operations conducted on enemy territory without a direct engagement between the military forces involved. To destroy the installations at *Al-Kibar*, a strike force of eight IAF warplanes could operate undetected in the heavily defended Syrian airspace from start to finish. To this end, the Israelis synergistically maneuvered various assets across the domains of military operations. Therefore, the operation provides the opportunity to revisit and extrapolate relative superiority theory to describe how this condition was achieved across domains and, secondly, how this condition enabled the generation of cross-dimensional effects.

Did the Israelis achieve relative superiority throughout the operation?

The Israelis achieved relative superiority across multiple domains of military operations. Figure 4.2 illustrates the ability of the Israelis to achieve relative superiority by significantly decreasing the area of vulnerability from the onset of the operation. This was mainly the result of superior Israeli and American intelligence, leading to identifying the site at *Al-Kibar* in the first place. Thus, the Israelis became aware of the site's existence, while the Syrians were unaware of that knowledge. Because they believed to be building the site undetected, the Syrians did not protect it with the appropriate defensive measures, such as anti-aircraft guns, which illustrated a critical vulnerability in the Syrian defenses. Furthermore, Israeli intelligence on the Syrian IADS allowed the IDF to prepare the vital EW capabilities required to manipulate these systems, which is the second critical Syrian vulnerability. In sum, the Israeli information advantage allowed them to set the desired conditions to exploit the Syrian vulnerabilities and launch a successful surprise attack.

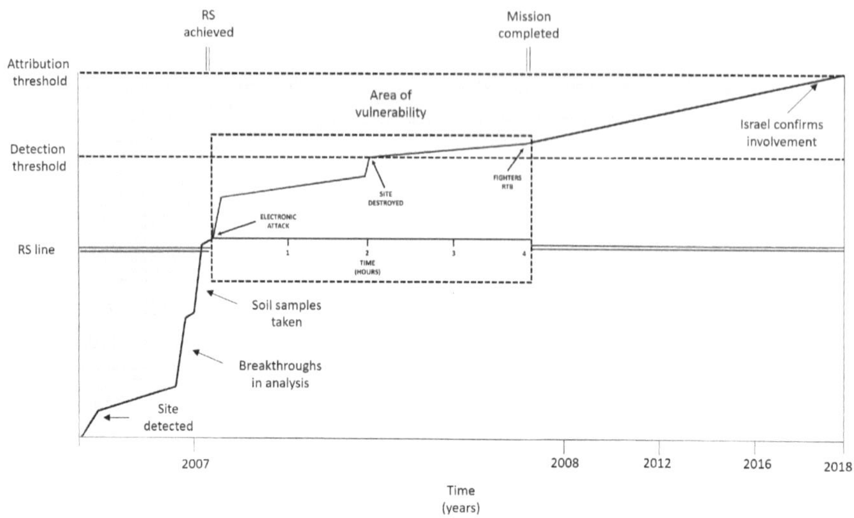

Figure 4.2 Relative superiority graph for the Israeli strike at *Al-Kibar*.

By conducting the EA at the beginning of the execution phase, the Israelis could gain complete control of the Syrian IADS and alter the imagery the Syrians saw on their screens by conducting a spoofing attack. This allowed the Israeli strike force to enter Syrian airspace undetected and achieve relative superiority. Thus, they reduced the area of vulnerability by half, as illustrated by Figure 4.2.

Having achieved relative superiority at this early stage in the operation meant that for the Israelis, the chance of mission success outweighed the chance of mission failure. Of course, the crews that were making their way to the target remained vulnerable between the moment relative superiority was achieved and mission completion. However, the crews mitigated their vulnerability by employing a variety of superior TTPs, including flying below 200 feet AGL and maintaining radio silence, thereby further decreasing the chances of detection.

If achieved, how did the Israelis sustain their relative superiority throughout the operation? If not, why did they fail?

After the successful infiltration, which took the crews approximately two hours, they destroyed the installations at *Al-Kibar* within minutes. At this point, the operation reached above the detection threshold, as the activities (i.e., the bombing of the site) or at least its effect (i.e., the destruction of the site) became visible to the Syrians. Importantly, the strike force itself was not detected, as the Syrian IADS was still manipulated. However, the mission was not completed by destroying the target: to successfully complete their mission, the crews had to return to base safely. The exfiltration meant flying another two hours in enemy airspace, posing significant threats to the strike force. The Israelis, therefore, had to sustain the condition of relative superiority until the fighters were back in Israelian airspace.

Sustaining relative superiority was achieved through the continued manipulation of the Syrian IADS and employing the TTPs mentioned above.

In sum, the Israelis achieved and sustained relative superiority because they had detailed and reliable intelligence, used modern technology, employed superior TTPs, and maintained excellent OPSEC throughout the whole operation. These factors combined allowed the Israelis to significantly reduce all risks involved so that the strike force could enter Syrian airspace, destroy the site, and safely return to base, all while remaining undetected.

Conclusion

As described in the introductory chapter of this book, policymakers are responsible for defining political objectives, while military commanders are responsible for employing tactics. Exploiting the full potential of strategic special operations is thus highly dependent on bilateral consultation and mutual understanding of strategy development and tactical behavior.

Operation Orchard illustrates how superior tactical action can secure political objectives. The Israelis accomplished this by considering the operation from a systems perspective. They synergistically employed various instruments of national power across the government, allowing them to achieve an information advantage over the Syrians. The brilliant tactical execution illustrates that achieving relative superiority remains a constant and critical condition for mission success. The potential ways of achieving this, however, have expanded, as the Israelis demonstrated the ability to achieve this condition across multiple domains of military operations (i.e., air, land, space, and the EMS), which allowed them to generate cross-dimensional effects (i.e., physical, virtual, and cognitive).

In conclusion, the Israeli policymakers and military commanders demonstrated a high level of strategic sense as they well understood that converting tactical performance into strategic effects required them to look at the nature of the whole operation, not its individual parts. Operation Orchard, therefore, illustrates one in which the Israelis fulfilled the promise of strategic special operations through the careful alignment of policy, strategy, and tactics.

Notes

1 General (Ret.) Michael Hayden cited in: Ronen Bergman, "Ex-CIA Director: I Was Sure If We Didn't Strike Syria's Nuclear Reactor, Israel Would," *Ynetnews*, December 30, 2016.
2 Erich Follath and Holger Stark, "The Story of 'Operation Orchard': How Israel Destroyed Syria's Al Kibar Nuclear Reactor," *Spiegel Online International*, November 2, 2009.
3 Follath and Stark, "The Story of 'Operation Orchard'."
4 Nicholas Blanford, "Lebanon: The Shiite Dimension," *Wilson Center*, July 13, 2022.
5 Arms Control Association, "Arms Control and Proliferation Profile: Syria," *Fact Sheets & Briefs*, last reviewed: March 2022.
6 Follath and Stark, "The Story of 'Operation Orchard'."
7 Hayden, cited in: Bergman, "Ex-CIA Director."

8 Senior Intelligence Official cited in: "Background Briefing with Senior U.S. Officials on Syria's Covert Nuclear Reactor and North Korea's Involvement," transcript, Office of the Director of National Intelligence, April 24, 2008.
9 Bergman, "Ex-CIA director."; Senior Intelligence Official cited in: "Background Briefing."
10 General Muhammad Suleiman was assassinated in Tartous, Syria on August 1, 2008 by Israeli SOF: BBC, "Israeli commandos' killed top Syrian official - leaked US files," July 16, 2015.
11 Hayden cited in: Bergman, "Ex-CIA director."
12 Federation of American Scientists, "A Sourcebook on the Israeli Strike in Syria," September 6, 2007.
13 Amnon Sufrin cited in: Amos Harel and Aluf Benn, "No Longer a Secret: How Israel Destroyed Syria's Nuclear Reactor," *Haaretz*, March 23, 2018.
14 General (Ret.) Amos Yadlin cited in: Harel and Benn, "No Longer a Secret."
15 Federation of American Scientists, "A Sourcebook."
16 Ibid."; Harel and Benn, "No Longer a Secret."
17 Federation of American Scientists, "A Sourcebook."
18 Ibid.
19 Rami Ben-Barak cited in: Harel and Benn, "No Longer a Secret."; Warren Bass, *A Surprise Out of Zion? Case Studies in Israel's Decisions on Whether to Alert the United States to Preemptive and Preventive Strikes, from Suez to the Syrian Nuclear Reactor* (Santa Monica, CA: RAND Corporation, 2015).
20 Yaakov Katz, *Shadow Strike: Inside Israel's Secret Mission to Eliminate Syrian Nuclear Power* (New York: St. Martin's Press, 2019), 16–17.
21 Bergman, "Ex-CIA Director"; Katz, *Shadow Strike*, 17; Senior Intelligence Official cited in: "Background Briefing."
22 Bass, *A Surprise Out of Zion?*, 47
23 Sufrin, cited in: Harel and Benn, "No Longer a Secret."
24 Israeli Prime Minister Ehud Olmert cited in: Bergman, "Ex-CIA Director."
25 George W. Bush, "The White House: President Bush's Statement on North Korea Nuclear Test," Office of the Press Secretary, October 9, 2006.
26 Meir Dagan cited in: Katz, *Shadow Strike*, 28.
27 Hayden cited in: Bergman, "Ex-CIA Director."
28 Bergman, "Ex-CIA Director."
29 Ibid.
30 U.S. Senior Official cited in: David Makovsky, "The Silent Strike: How Israel Bombed a Syrian Nuclear Installation and Kept It Secret," *The New Yorker*, September 17, 2012.
31 Bergman, "Ex-CIA director."
32 Senior Intelligence Official cited in: "Background Briefing."
33 Senior Intelligence Official cited in: "Background Briefing."
34 Bergman, "Ex-CIA director."
35 Michael V. Hayden, *Playing to the Edge: American Intelligence in the Age of Terror* (New York: Penguin Press, 2016), 258.
36 Bergman, "Ex-CIA director."
37 Harel and Benn, "No Longer a Secret."
38 Ibid.; Bass, *A Surprise Out of Zion?*
39 Dan Raviv and Yossi Melman, *Spies Against Armageddon: Inside Israel's Secret Wars* (Sea Cliff, NY: Levant Books, 2012), 315; Katz, *Shadow Strike*, 159–163.
40 Makovsky, "The Silent Strike."
41 Ibid.
42 Katz, *Shadow Strike*, 20.
43 Senior Intelligence Official cited in: "Background Briefing."
44 Sufrin cited in: Harel and Benn, "No Longer a Secret."

45 Dick Cheney, *In My Time* (New York: Threshold Editions, 2011) 465–466.
46 Senior Intelligence Official, cited in: "Background Briefing."
47 Harel and Benn, "No Longer a Secret."
48 General Ido Nehoshtan cited in: Harel and Benn, "No Longer a Secret."
49 Harel and Benn, "No Longer a Secret."
50 Colonel Amir cited in: Harel and Benn, "No Longer a Secret."
51 Ibid.
52 Ibid.
53 Ibid.
54 Ibid.
55 Follath and Stark, "The Story of 'Operation Orchard'."
56 Colonel Amir cited in: Harel and Benn, "No Longer a Secret."
57 Lior Tabansky and Isaac Ben Israel, "Striking with Bits? The IDF and Cyber-Warfare," in *Cybersecurity in Israel. SpringerBriefs in Cybersecurity* (Springer, 2015), 65.
58 David A. Fulghum, Robert Wall, and Amy Butler, "Israel Shows Electronic Prowess," *Aviation Week & Space Technology* 168, no. 25 (2007): 29–30; David A. Fulghum, Robert Wall, and Amy Butler, "Cyber-Combat's First Shot," *Aviation Week & Space Technology* 167, no. 16 (2007): 28–31; Sally Adee, "The Hunt for the Kill Switch," *IEEE Spectrum* 45, no. 5 (2008): 34–39.
59 Fulghum, Wall, and Butler, "Israel Shows Electronic Prowess," 29–30; Adee, "The Hunt," 34–39.
60 John Toepher, "SEAD from the Ground up: SOF's Role in the Suppression of Enemy Air Defenses," (Master thesis, Naval Postgraduate School, 2019), 39.
61 Fulghum, Wall, and Butler, "Cyber-Combat's First Shot," 29–30.
62 Makovsky, "The Silent Strike."
63 Sławomir Dygnatowski, Paweł Dygnatowski, and Łukasz Domżał-Drzewicki, "The analysis of using structural solutions in cybersecurity based on Orchard Operation," *Journal of KONBiN* 49 (2019): 281–298; Katz, *Shadow Strike*, 186.
64 Adee, "The Hunt for the Kill Switch."
65 Ibid., 34.
66 Ibid., 36.
67 Ibid., 38.
68 Ibid., 38.
69 Raviv and Melman, *Spies Against Armageddon*, 159–163.
70 Tabansky and Israel, Striking with Bits?, 65.
71 Colonel Amir cited in: Harel and Benn, "No Longer a Secret,"
72 Ibid.
73 Ibid.
74 Ibid.
75 BBC News, "Syria Fires on Israel Warplanes," September 6, 2007.
76 Ibid.
77 Yoav Stern and Mazal Mualem, "Israel Mum on Any IAF Entry into Syria Airspace," *Haaretz*, September 7, 2007.
78 Reuters, "Israel Didn't Attack Syria: Israel Radio," September 6, 2007.
79 Yaakov Katz, *Shadow Strike*, 243.
80 Mark Mazzetti and Helene Cooper, "U.S. Confirms Israeli Strikes Hit Syrian Target Last Week," *New York Times*, September 12, 2007.
81 Stephen Farrell, "Israel Admits Bombing Suspected Syrian Nuclear Reactor in 2007, Warns Iran," *Reuters*, March 20, 2018.
82 Israeli Prime Minister Benjamin Netanyahu, Tweet, March 21, 2018.
83 Colonel Amir cited in: Harel and Benn, "No Longer a Secret."
84 Senior Intelligence Official cited in: "Background Briefing."
85 Farrell, "Israel Admits."
86 Reuters, "Israel Didn't Attack Syria."

Bibliography

Adee, Sally. "The Hunt for the Kill Switch," *IEEE Spectrum* 45, 5 (2008).

Arms Control Association. "Arms Control and Proliferation Profile: Syria," *Fact Sheets & Briefs*, March 2022. https://www.armscontrol.org/factsheets/syriaprofile.

"Background Briefing with Senior US Officials on Syria's Covert Nuclear Reactor and North Korea's Involvement (April 24, 2008). https://www.dni.gov/files/documents/Newsroom/Speeches%20and%20Interviews/20080424_interview.pdf.

Bass, Warren. *A Surprise Out of Zion? Case Studies in Israel's Decisions on Whether to Alert the United States to Preemptive and Preventive Strikes, from Suez to the Syrian Nuclear Reactor*, (Santa Monica, CA: RAND Corporation, 2015), https://www.rand.org/pubs/research_reports/RR498.html.

BBC. "Israeli Commandos' Killed Top Syrian Official - Leaked US Files," July 16, 2015. https://www.bbc.com/news/world-middle-east-33548065.

BBC News. "Syria 'Fires on Israel Warplanes," September 6, 2007, http://news.bbc.co.uk/2/hi/middle_east/6981674.stm.

Bergman, Ronen. "Ex-CIA Director: I Was Sure If We Didn't Strike Syria's Nuclear Reactor, Israel Would," *Ynetnews*, December 30, 2016. https://www.ynetnews.com/articles/0,7340,L-4900816,00.html.

Blanford, Nicholas. "Lebanon: The Shiite Dimension," *Wilson Center*, July 13, 2022. https://www.wilsoncenter.org/article/lebanon-the-shiite-dimension.

Bush, George W. "The White House: President Bush's Statement on North Korea Nuclear Test," Office of the Press Secretary, October 9, 2006, https://georgewbush-whitehouse.archives.gov/news/releases/2006/10/20061009.html#:~:text=THE%20PRESIDENT%3A%20Last%20night%20the,States%20condemns%20this%20provocative%20act.

Cheney, Dick. *In My Time* (New York: Threshold Editions, 2011).

Clausewitz, von Carl. *On War*, ed. and trans. by Michael Howard, Peter Paret, and Bernard Brodie (Princeton, N.J.: Princeton University Press, 1984).

Dygnatowski, Sławomir, Pawel Dygnatowski, and Łukasz Domżał-Drzewicki. "The Analysis of Using Structural Solutions in Cybersecurity Based on Orchard Operation," *Journal of KONBiN* 49 (2019): 281–298.

Federation of American Scientists. "A Sourcebook on the Israeli Strike in Syria," September 6, 2007, https://man.fas.org/eprint/syria.pdf.

Follath, Erich, and Holger Stark. "The Story of 'Operation Orchard': How Israel Destroyed Syria's Al Kibar Nuclear Reactor," *Spiegel Online International*, November 2, 2009, https://www.spiegel.de/international/world/the-story-of-operation-orchard-how-israel-destroyed-syria-s-al-kibar-nuclear-reactor-a-658663.html.

Fulghum, David, Robert Wall, and Amy Butler. "Cyber-Combat's First Shot," *Aviation Week & Space Technology* 167 (2007).

Fulghum, David, Robert Wall, and Amy Butler. "Israel Shows Electronic Prowess," *Aviation Week & Space Technology* 168 (2007).

Harel, Amos, and Aluf Aluf Benn. "No Longer a Secret: How Israel Destroyed Syria's Nuclear Reactor," *Haaretz*, March 23, 2018, https://www.haaretz.com/world-news/2018-03-23/ty-article-magazine/no-longer-a-secret-how-israel-destroyed-syrias-nuclear-reactor/0000017f-e518-d9aa-afff-fd58293e0000.

Hayden, Michael V. *Playing to the Edge: American Intelligence in the Age of Terror* (New York: Penguin Press, New York, 2016).

Katz, Yaakov, and *Shadow Strike. Inside Israel's Secret Mission to Eliminate Syrian Nuclear Power* (New York: St. Martin's Press, 2019).

Makovsky, David. "The Silent Strike: How Israel Bombed a Syrian Nuclear Installation and Kept It Secret," *The New Yorker*, September 17, 2012, https://www.newyorker.com/magazine/2012/09/17/the-silent-strike.

Mazzetti, Mark, and Helene Cooper. "US Confirms Israeli Strikes Hit Syrian Target Last Week," *The New York Times*, September 12, 2007, http://www.nytimes.com/2007/09/12/world/middleeast/12syria.html.

Netanyahu, Benjamin. Tweet, March 21, 2018, https://twitter.com/israelipm/status/976478189466869760.

Raviv, Dan, and Yossi Yossi Melman. *Spies Against Armageddon: Inside Israel's Secret Wars* (Sea Cliff, NY: Levant Books, 2012).

Reuters. "Israel didn't attack Syria: Israel Radio," *Reuters*, September 6, 2007, http://www.reuters.com/article/topNews/idUSL0679478220070906.

Stern, Yoav, and Mazal Mazal Mualem. "Israel mum on any IAF entry into Syria airspace," *Haaretz*, September 7, 2007, http://www.haaretz.com/hasen/spages/901650.html.

Tabansky, Lior, and Isaac Ben Israel. "Striking with Bits? The IDF and Cyber-Warfare," in: *Cybersecurity in Israel. SpringerBriefs in Cybersecurity* (Springer, 2015).

Toepher, John. "SEAD from the ground up: SOF's role in the suppression of enemy air defenses," (Master Thesis, Naval Postgraduate School, 2019).

US Department of Defense. *Joint Publication 3-0 Joint Operations* (Washington: GPO, 2017).

5 Operation Olympic Games

The Cyberattack at Natanz

Introduction

The previous chapter presented a direct special operation known as a preventive strike. These special operations are traditionally conducted by air, land, or sea. The planning and preparation phases to find and fix targets could occur over weeks, months, or even years. By contrast, the execution phase (i.e., the finish), due to its kinetic character, ranges from minutes to hours, depending on the precision-strike capability. Finally, Special Operations Forces (SOF) commonly play a substantial role in these operations.

This chapter presents another example of a direct special operation with a similar purpose: a preventive strike. However, the operation's design is totally different from the traditional ones conducted by air, land, or sea. In this case, the emergence of cyberspace allowed for a strategy in which the United States (US) National Security Agency (NSA) and Israeli Unit 8200 (from now on, "the attackers"), who allegedly conducted the operation, took advantage of the unique characteristics of this new domain of military operations. This allowed them to strike the target just by electronic means and from across the globe while significantly decreasing the risk to the force, as it was almost exclusively a virtual one. Finally, the case study illustrates that by conducting special operations through cyberspace, the execution phase of a precision strike could take place over an extended period of time. In the case of Operation Olympic Games over multiple years.

Background

On August 19, 1953, the government of Premier Mohammad Mosaddeq was overthrown. Although conducted by the Iranian military, the coup was widely seen as a covert operation conducted by the US Central Intelligence Agency (CIA). Immediately upon the removal of Mosaddeq, Shah Mohammed Reza Pahlavi was reinstated as Iran's leader and served as an ally to the US and the United Kingdom (UK).

In 1957, the US wanted to increase its military and economic assistance to Iran and started actively supporting Iranian efforts to develop nuclear energy. On March 5, 1957 (under the Eisenhower Atoms for Peace Program), both countries publicly announced a "proposed agreement for cooperation in research in the peaceful uses of atomic energy."[1] The agreement allowed the US to increase its

DOI: 10.4324/9781003365112-7

footprint in Iran's civilian nuclear industry, including the healthcare and medicine markets.[2] US investments included the lease of 13.2 pounds of Low-Enriched Uranium (LEU) provided by the US Atomic Energy Commission.[3]

Two years after the agreement, the Shah announced the establishment of the Tehran Nuclear Research Center. He successfully persuaded the US to provide the Iranians with a five-megawatt reactor. Iran's intentions with the nuclear program remained transparent throughout the next decade. When the Nuclear Nonproliferation Treaty (NPT) opened for signature in 1968, the Iranians immediately signed and ratified it two years later, thereby granting the International Atomic Energy Agency (IAEA) permission to verify and safeguard its nuclear program.[4]

Iran completed the Safeguards Agreement with the IAEA by 1974.[5] As a result, international support for the Iranian nuclear program grew, and within the next couple of years, France and Germany joined the US in its cooperation with Tehran. At the same time, the Shah approved the establishment of the Atomic Energy Organization of Iran (AEOI) and clarified that Iran aimed to produce 23,000 megawatts of electricity through its nuclear energy program. Iran signed contracts with French and German private firms to construct nuclear sites and receive nuclear fuel supplies to achieve that goal.[6]

Despite Iran's membership of the NPT, which implies that the IAEA safeguarded its nuclear installations, international support for its nuclear program started to degrade in August 1974. During that time, a US Special National Intelligence Estimate assessed the impact of India's successful nuclear test on Iran's nuclear policy.[7] The CIA estimated that Iran would see the Indian test as a significant threat to its national security and might persuade Tehran to start developing its own nuclear weapons program:

> There is no doubt, however, of the Shah's ambition to make Iran a power to be reckoned with. If he is alive in the mid-1980s, if Iran has a full-fledged nuclear power industry and all the facilities necessary for nuclear weapons, and if other countries have proceeded with weapons development, we have no doubt that Iran will follow suit.[8]

The intelligence assessment resulted in the withdrawal of Western support for the Iranian nuclear program, including the cancellation of ongoing construction efforts to build nuclear reactors by France and Germany.[9] The Islamic Revolution in 1979 that toppled the Shah and included the Iran hostage crisis, during which supporters of Iran's Supreme Leader, Ayatollah Ruhollah Khomeini, seized the US Embassy in Tehran and took 52 American diplomats and citizens hostage, led the US to decide to terminate its diplomatic relations with Tehran. This effectively meant the end of Western support for the Iranian nuclear program. Shortly after that, Iraq invaded Iran, which marked the start of a protracted war of attrition in which an estimated 1,000,000 people were killed. Throughout the war, Iraq received substantial military support from the US. Iraq also developed nuclear capabilities that posed an imminent threat to Iran's national interests, including regime security.

This perceived threat led to the Iranian desire to establish nuclear security, as the regime believed that regime stability went hand-in-hand with the possession of nuclear weapons. To this end, Khomeini sought to develop new partnerships with Argentina, China, and Russia to complete the unfinished construction efforts in response to these developments. These attempts were unsuccessful, as US sanctions and impediments forced these countries to withdraw from the projects, leaving the Iranians to complete them on their own. Mohammad Javad Zarif, the former Iranian ambassador to the United Nations (UN), argued that by doing so, the US left the Iranian government with no other choice than to continue its nuclear program clandestinely:

> Iran was left with no option but to be discrete in its peaceful activities. In doing so, Iran remained within the confines of the NPT and did not divert its peaceful program to military activities. Meanwhile, in order to avoid the US-lead restrictions and impediments, Iran refrained from disclosing the details of its programs, which in nearly all cases—including the most publicized cases of the Natanz enrichment facility and the Arak heavy water plant—it was not obliged to disclose under the terms of its existing safeguards agreement with the IAEA.[10]

The immediate consequences of the Western withdrawal and US sanctions delayed the Iranian nuclear program. Over time, several other factors drove the Iranian program into a decline. For example, Khomeini opposed the nuclear program, and Iran was also confronted with an exodus of nuclear scientists.

In the mid-1980s, it appeared that Iran was making new attempts to develop its nuclear program. Leonard Spector, deputy director of the Center for Nonproliferation Studies, argues that around 1985, Iran started to receive assistance from Abdul Qadeer Khan, a Pakistani nuclear scientist, in contacting suppliers of nuclear technology.[11] The decisive factor in rebuilding Iran's nuclear program, however, was the death of Khomeini in 1989, which paved the way for the appointment of Ayatollah Ali Khamenei, who viewed the program more positively, as Iran's new Supreme Leader. Other factors that might have persuaded the Iranians to boost its nuclear program might have been Iraq's secret nuclear weapons program, which was revealed during the 1991 Gulf War, and the US its increasing footprint in the Middle East.

Over the next two decades, Iran received assistance from China, Russia, and North Korea to finish the construction of its nuclear facilities at Bushehr and Darkhovin. In addition to these two sites, Iran operates a variety of installations, some of which are considered major nuclear sites and are suspected of being critical to Iran's nuclear weapons program. For example, the US Congressional Research Service argued that the Isfahan Nuclear Technology Center is a principal site for uranium conversion and reportedly has produced approximately 366 tons of uranium hexafluoride.[12] The Iranians use converted uranium, also known as uranium feedstock, to be fed into another nuclear site that serves as a critical component in its nuclear arms program: the enrichment facility at Natanz.

By the early 2000s, the US believed that Iran posed a severe threat to global security. A nuclear weapon in the hands of the regime in Tehran could be disastrous. In 2002, US President George W. Bush stated that Iran, Iraq, and North Korea formed the so-called "Axis of Evil," arguing that:

> States like these, and their terrorist allies, constitute an axis of evil, arming to threaten the peace of the world. By seeking weapons of mass destruction, these regimes pose a grave and growing danger. They could provide these arms to terrorists, giving them the means to match their hatred. They could attack our allies or attempt to blackmail the United States. In any of these cases, the price of indifference would be catastrophic.[13]

Over the next couple of years, most US strategic documents referred to the threat posed by Iran, including nuclear and missile programs.[14] Diplomatic pressure, United Nations Security Council (UNSC) Resolutions calling on Iran to stop the development of its nuclear program without delay, and subsequent economic sanctions imposed by the US were ineffective.[15] Due to the ineffectiveness of these coercive policies, the US started to develop plans to apply military force against Iran, including special operations raids.[16] Israel also pushed the US to use military force against Iran, just as in 2007 when former Israeli Prime Minister Ehud Olmert had tried to persuade President Bush to strike against a Syrian nuclear site known as *Al-Kibar* (see Chapter 4).[17] Given that the conduct of such a strike on Iranian soil would be much riskier, Bush treated this option as a last resort.[18]

In addition to the increased risk of a potential US strike on Iranian soil, US intelligence assessed that the Iranians might have terminated their nuclear weapons program in 2003. The intelligence report, Bush argued, undermined the ongoing diplomatic efforts and prevented him from steering toward a military option.[19] Bush believed that the intelligence assessment might have resulted from the CIA's caution not to make the same mistakes as they did in the Iraqi case in 2003.[20] Secretary of Defense Robert Gates provided further skepticism about the intelligence estimate and found "that the report had presented the evidence poorly, underemphasizing the importance of Iran's enrichment activity and overemphasizing the suspension of a weapons-design effort that could easily be turned back on."[21] Notably, almost simultaneous with the release of the intelligence estimate, the Israeli press published a news story about Iran's intensification of its nuclear program.[22]

Whether Iran was increasing or decreasing its nuclear program remains unclear. Despite opposing views, Israel requested US assistance in conducting an air strike on the Natanz nuclear complex. The request included delivering a particular munition developed to strike targets located deep underground, such as the underground facilities at Natanz. Furthermore, Israel requested the fly-over rights of navigable Iraqi airspace and refueling capabilities that would enable the Israeli warplanes to increase their operational reach required for the strike. The US declined the Israeli requests.[23]

Simultaneously, though, the US offered Israel an alternative. This alternative intended to delay the Iranian nuclear program rather than destroy the site at Natanz. The approach was focused on manipulating Natanz's critical infrastructure, including the operation of its centrifuges. The US had previously employed such an approach by delivering defective power supply units, resulting in an explosion and the destruction of approximately 50 centrifuges.[24] This time, however, the US was developing a much more sophisticated plan for sabotage that centered on using malicious software (malware).

The Natanz enrichment complex

On August 14, 2002, the National Council of Resistance of Iran (NCRI), an organization that is home to Iranian opposition factions in exile, made public that the regime in Tehran was building a nuclear fuel manufacturing plant near Natanz. This ancient town is situated approximately 360 kilometers south of the Iranian capital of Tehran. During the press conference, Alireza Jafarzadeh, the NCRI's spokesperson, revealed that:

> Although on the surface, [Iran's] main nuclear activity revolves around [the] Bushehr nuclear plant .... In reality, many secret nuclear programs are at work without any knowledge of [the] International Atomic Energy Agency .... Today, I am going to reveal to you two top-secret sites of the Iranian regime that they have succeeded to keep secret until today.[25]

Jafarzadeh argued that the site's development was the responsibility of Iran's AEOI in cooperation with its Supreme National Security Council. It was a clandestine project disguised by several front companies that could secretly procure materials and technology for its development. The enrichment complex covers 100,000 square meters and is situated along a highway, which links Natanz with the city of Kashan. According to Jafarzadeh, the site's construction began in 2000 and was approximately 90 days from completion. Upon completion, the required equipment would be installed, and, in that regard, the site was nearing operational capability.

To maintain the project's clandestine character, the Iranians developed a cover story that suggested the site was a desert eradication project. For several reasons, the cover story was relatively weak. For example, a former Iranian Prime Minister, now a Supreme National Security Council representative, visited the site in July 2002. According to Iranian norms, such a high-level visit to a desert eradication project would be inappropriate. Suspicion about the true nature of the site grew further after it became known that the head of the AEOI frequently visited the nearby town Kashan, likely to be kept updated on the construction efforts. Finally, construction workers were not allowed to share details of their work with local officials, and the AEOI would not share details of the project with the Office of the Governor of Kashan. Friction between the local authorities and the AEOI increased when the Deputy Governor General of Kashan Province did not receive permission to visit the site and was denied access.[26]

Whether the information presented by the NCRI was coming from Iranian sources inside Iran or was provided by US or Israeli Intelligence remains to be determined. The CIA, however, was aware of the Natanz site before the NCRI press conference, as emphasized by then-director George Tenet: "I want to assure you that recent Iranian admissions about their nuclear programs validate our intelligence assessment. It is flat wrong to say we were 'surprised' by reports from the Iranian opposition last year."[27]

Importantly, the US State Department had designated the NCRI a foreign terrorist organization in 1997 because it was either part of or supported the Mujahedin-e Khalq (MEK), a group notorious for its opposition to Israeli and US foreign policy in the Middle East.[28] By exposing the clandestine project at Natanz, the NCRI gained support from several US lawmakers, as well as former senior officials of the CIA and Federal Bureau of Investigation (FBI).

With the information provided by the NCRI, Corey Hinderstein of the Institute for Science and International Security identified the site from commercial satellite imagery. When Hinderstein was searching for the location by using Digital Globe, several signs on the map suggested that others had already identified the site and had been looking for information. From the images, Hinderstein could identify several buildings and large mounds of churned earth protected by a double security perimeter. The fact that there was only one road leading to the site suggested that access to and from the site could be easily controlled. In the direct vicinity of the buildings, several circles were identified that would likely mark the future positions of anti-aircraft guns. These were significant findings, as the identification of Soviet nuclear weapons in Cuba in 1962 was an important discovery that sparked the beginning of the Cuban missile crisis. An overview of the Natanz enrichment complex is illustrated in Figure 5.1.

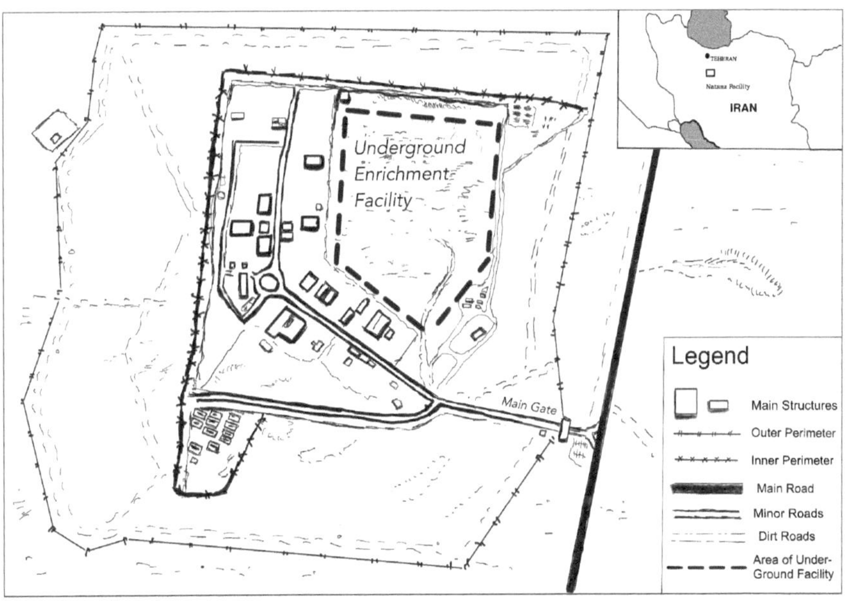

Figure 5.1 Overview of the Natanz enrichment complex.

Most interesting were the images that revealed three large buildings and a connecting tunnel built deep underground, two of which were identified as production halls. The walls of these buildings were approximately six to eight feet thick, indicating they would serve as protective measures against air strikes. This suggestion was supported by the U-shape of the tunnel leading down to the buildings, which is a common design to protect underground facilities against missile strikes. The three buildings were home to a vital part of the facility, namely the centrifuges, a critical component of the uranium enrichment program. The two production halls were large enough to hold 50,000 centrifuges. By November 2009, however, reports estimated that the Iranians had been able to install approximately 8,700 centrifuges.[29]

Planning and preparation

By late 2007, US President George W. Bush reserved $400 million to fund activities that would undermine Iran's nuclear program. Thus, allowing intelligence collection, political warfare, and covert action to sabotage the facilities and equipment used by the Iranians.[30] The covert action component was designated explicitly for offensive cyber operations that would, among other things, sabotage the computer systems at Natanz to delay the Iranian nuclear program. However, the designation of this funding in 2007 does not appear to signal the start of the planning and preparation of the operation. Indeed, several features in the malware (officially known as W32.Stuxnet code or Stuxnet) indicated that the planning and preparation phases had already begun several years prior.[31] For example, the Industrial Control Systems (ICSs) at Natanz are commanded and controlled by Programmable Logic Controllers (PLCs), which contain a type of computer code similar to assembly code.[32] The PLCs are usually programmed through Windows software that runs on stand-alone computers disconnected from the Internet and air-gapped networks. The ICSs are also in stand-alone modus during the programming process. As a result, the attackers had to start with the first step in their attack, reconnoitering the site at Natanz and looking for vulnerabilities that they could exploit.

During this reconnaissance phase, vulnerabilities were successfully exploited, closely followed by the delivery phase. In the case of Operation Olympic Games, a helpful first step would have been to collect information about the schematics of the S7-315 and S7-417 PLCs used at Natanz because they were configured uniquely.[33] As the execution phase of the operation has indicated, the attackers successfully collected the required information to develop the destructive payload. They determined the most effective way to go through the other stages of the attack, including delivery, installation, and lateral movement, while remaining undetected. The exact timeframe during which the attackers were able to collect this critical information remains to be determined, as a variety of indicators point in different directions.

For example, the timestamps in the code through which Stuxnet infected the PLCs S7-315 and S7-417 indicate that the binaries were compiled in 2000 and 2001. Furthermore, the timestamp that was found in the malicious .DLL (dynamically linked libraries) file, which allowed the malware to replace the original Siemens SIMATEC Step Step 7 .DDL, has indicated it had been compiled in 2003.

On the one hand, the three timestamps could be legitimate, meaning that the attackers kept the code on the shelf until it was the appropriate time to employ it. On the other hand, the attackers could have manipulated the timestamps to deceive the Iranians or other actors in case Stuxnet was exposed.

The fact that specific code blocks were targeting the valves and frequency converters of the centrifuges that were specifically tailored to the configuration at Natanz supports the argument of deception (i.e., forgery of the timestamps). In this case, the attackers should have been able to know with foresight which specific components would be installed in the facility, as well as knowledge about their configuration and modes of operation. Having such a detailed understanding at the time of the timestamp seems unlikely, as the facility was not completed yet. CIA case officers, however, allegedly had recruited several members of the Pakistani Abdul Qadeer Khan network who could have given them information about the specific parts Iran was acquiring, including the Siemens equipment. According to David Albright, most of the information about the components, configuration, and operations could have been known as early as 2001, but doubts whether the same applies to the frequency converters as their origin would have been uncertain at that time:

> The cascade details, including the 164 centrifuges per cascade, number of stages [in the cascade], most valves, pressure transducers, and piping, could have been known [that early]. Frequency converters would be another matter, since Iran was acquiring them abroad back in that period from a variety of companies. So it would be hard to believe that Stuxnet's designers in 2001 could count on them being from Finland or domestically assembled [by Fararo Paya]. Moreover, the first module [of cascades installed at Natanz in 2007] was built with a range of important frequency converters.[34]

The timestamp with a compilation date of 2003 found in the Step 7 .DDL indicates a more plausible date, as more information about the Natanz site was available by that time. Indeed, it was in February 2003 when IAEA inspectors visited the Natanz site for the first time. By then, Iran had already installed a small cascade at the facility and planned to install approximately 1,000 centrifuges by the end of 2003.

The IAEA visit required Iran to provide the inspectors with an overview of the materials procured for its nuclear facilities, including the site at Natanz. In addition, Western intelligence agencies were also collecting information on Iran's secret procurement activities, including by operating through a front company named Neda Industrial Group, based in Tehran. Neda worked closely with Kalaye Electric, which had changed its business model from producing watches to centrifuges. Neda had previously partnered with Siemens and installed Siemens S7 PLCs (the same model Stuxnet attacked) in other Iranian facilities. With this information available to the attackers, they could have assumed that the same type of PLCs would also be installed at Natanz. Investigative journalist Kim Zetter states that she was given access to a letter confirming Iran had purchased Siemens equipment in 2002 and 2003.[35] Despite these findings, it remains uncertain when the various codes were compiled.

What can be said with certainty is that the code was highly sophisticated. First, the size of Stuxnet was 22 times larger than other types of malware found at that time. Second, the malicious binaries contained driver files that were digitally signed with two stolen certificates from JMicron and Realtek, which were legitimate. Third, the code included four zero-day Windows exploits and an additional three zero-day exploits in the PLC software, which is unique and indicates the significant investments made in developing the code.[36] Because of its sophistication, security researchers at Symantec assessed that the code's development would have required unique programming capabilities employed by five to ten people over six months while having access to Supervisory Control and Data Acquisition (SCADA) systems.[37]

To develop the Stuxnet code, its developers would have required detailed intelligence on the components, configuration, and mode of operation of the Natanz facility. For example, they had to know at what specific frequency the Iranians operated the converters. The attackers were also required to have a detailed understanding of the Step 7 system and the computer network at the facility to prevent cascading effects into its other systems. A crucial question in that regard was whether the facility's network was so-called "air-gapped." In other words, was the network at Natanz isolated, or were some of its parts connected with the outside world? In the latter case, spreading Stuxnet would affect the computers of other organizations or even private citizens, thereby causing potential damage. More importantly, it would alert governments or private companies who, in turn, could expose the operation and thereby lose its clandestine character.

Exactly how the attackers collected the information about the network at Natanz remains uncertain. Zetter argues that the information could have been collected by using advanced virtual spying tools like Flame and Duqu.[38] However, similar to the question about the delivery mechanism for Stuxnet, these two exploits should have been delivered either through the Internet (for example, via a phishing attack) or by using a USB flash drive to bridge the air gap. Next to a flash drive, the malware could have been installed in a water pump, which, once delivered to Natanz, was connected to the Local Area Network (LAN). David Sanger and Thom Sanker offer a third possibility: data about infected air-gapped systems could be transmitted by using electronic waves in the electromagnetic spectrum (EMS).[39]

Reportedly, the attackers used Erik van Sabben, a Dutch agent whom the Dutch intelligence service AIVD (Algemene Inlichtingen en Veiligheidsdienst) had recruited, to accomplish this.[40] The AIVD recruited him at the request of the CIA and Mossad.[41] To access the Natanz installation, van Sabben used his position within TTS International, a Dubai-based company that delivered parts to Iran's oil and gas companies, to access the site.[42] He successfully infiltrated the Natanz site in 2007 and provided the AIVD with critical information regarding the components and configuration of the control system at Natanz that greatly supported the development of the Stuxnet code.[43]

The protracted character of the planning and preparation phases indicates their significance. Indeed, there were many contingencies that the attackers had to consider because many things could go wrong once the malware was injected into the Natanz systems. In that regard, rehearsing the malware's mode of operation, including manipulating the Step 7 system or PLCs, played a vital role in eliminating potential barriers to success. Moreover, given that multiple versions of the Stuxnet code (i.e., version 0.5 and 1.001) were identified by the Symantec team indicates that the attackers rigorously tested how these differences in code would affect the centrifuges without alarming the Iranians that their systems were sabotaged. Indeed, the operation did not intend to destroy the centrifuges through brute force but rather slowly but surely while remaining undetected.

Similar to offensive actions that employ physical capabilities, conducting full-dressed rehearsals before a mission that uses virtual capabilities is vital to its success. Conducting a full-dressed rehearsal for a mission that is executed with virtual capabilities is challenging, as it requires a replica of the target's systems and network to ensure the malware's (or other attack vector's) mode of operation can be analyzed accordingly. In the case of Stuxnet, this task was extremely challenging as it required the attackers to build a replica of a Uranium-enriching nuclear installation to test the malware while preventing the exposure of such a significant part of the operation. An important question was: How could the attackers access the materials required to construct the replica? This is where the Oak Ridge National Laboratory comes into play.

The Oak Ridge National Laboratory is situated near Knoxville, Tennessee, US The site is managed by a nonprofit organization named UT-Battelle and founded by the Battelle Memorial Institute and the University of Tennessee. The site has a scientific character, and its research focuses, among other things, on nuclear science. More relevant to the operation are the classified national security studies that one of its laboratories conducts for the US Department of Defense (DOD), the US Department of Energy (DOE), and the US intelligence community. A replica of the Natanz site was built in a remote forested area of the Oak Ridge Reservation.[44] The replica initially served the purpose of understanding the Iranian capabilities and assessing the advancement of its nuclear program. In January 2004, however, the US was able to retrieve a variety of components of Libya's nuclear weapons program after former Libyan leader Moammar Gadhafi had struck a deal with the UK and the US in December 2003 to end two decades of international isolation and US sanctions.[45] The components included P-1 and P-2 centrifuges, similar to the ones the Iranians installed at Natanz.

With these components now available to the attackers, the malware could be extensively tested in a controlled environment. According to the Symantec team, attacking the PLCs required communication between the malware and the frequency converters.[46] The team found that the code used for this communication was modified in May 2006, corresponding to the date other essential malware functions were modified.[47] These modifications indicate the significance of the full-dressed rehearsals conducted at Oak Ridge and allowed the attackers to develop a digital weapon that was ready for use.

The attack

With the reconnaissance phase completed, the attackers could start with the actual attack on the site at Natanz. The attackers had to get the malicious code onto the site's control systems to attack the centrifuges. Because the site's network was air-gapped, the attackers could not deliver the malicious code via an attack method that required an Internet connection. Instead, the attackers had to bring the code physically into the facility and deliver it to its system. Sergey Ulasen, a cyber security analyst who then worked for VirusBlokAda, a small Belarussian antivirus software developer, offered an initial assessment of the Stuxnet code, indicating that the malware used a USB flash drive for propagation.[48] The USB flash drive (or the previously mentioned water pump) was brought into the Natanz site by van Sabben, who was recruited by the Dutch AIVD and had provided them with crucial information about the site's control system.[49] The number of deliveries that could be made with the USB flash drive was limited to three, and each delivery ran only for 21 days on the system. These limitations were likely added deliberately by the developers and indicate the attacker's intention to prevent the spread of the malware, decreasing the chance of detection and limiting collateral damage.

Once the USB flash drive had been brought into the Natanz facility and connected to one of the control system's computers, the delivery phase was completed. Because operating systems are typically designed to prevent the execution of unauthorized code, delivery by itself was insufficient, but its execution was also required.[50] For this, the malicious code had to be dropped onto the computer's hard disk to enable its installation. This is where the mrxnet.sys and mrxcls.sys drivers come into play, which exploited the Windows Operating System (OS) and injected the malicious code into the system's processes while simultaneously hiding the malware itself. This exploitation of the Windows OS was successful because both drivers used genuine digital certificates that were digitally signed by Realtek Semiconductor Corp, which allowed the malware to bypass automated detection programs.

With Stuxnet successfully executed, the malware could now control the first LAN computer. However, this computer was not the device the malware needed to start manipulating the centrifuges. To find the appropriate components, Stuxnet had to spread across the LAN to find and subsequently infect the components. This tactic is called lateral movement. More specifically, Stuxnet moved laterally across the LAN in search of the Field PGs (i.e., programmers), which are typically computers that run on Windows and were used to program Siemens PLCs.[51] Stuxnet would attempt to spread across the site's network by infecting other LAN computers by exploiting a zero-day vulnerability or through removable drives.[52] The Symantec team assessed that "propagation through a LAN likely served as the first step and propagation through removable drives [was the] means to cover the last and final hop to a Field PG that is never connected to an untrusted network."[53] A significant feature of Stuxnet was that all its capabilities were embedded directly in its executable program since the network at Natanz was air-gapped. In other words, the attackers could not control Stuxnet through a command-and-control server, a common tactic employed in digital attacks, but rather made Stuxnet self-operating. Once Stuxnet found a computer that ran Siemens SIMATEC Step 7 .DDL, the malware would automatically modify its malicious code on the PLC.[54] Modifying the PLCs is what likely caused the centrifuges to spin out of control and ultimately break them apart.

Stuxnet's mode of operation was highly sophisticated. The initial version (i.e., Stuxnet 0.5) targeted the valves of the centrifuges. The attack did not start immediately upon infection, but the malware observed and collected data about the centrifuges' mode of operation. Once the attack began, the malware only randomly targeted a limited number of centrifuges for a short duration (i.e., two hours). Each attack stage lasted approximately two hours, during which the selected valves were closed. This led to pressure build-up in the centrifuges and damaged them slowly but surely without alerting the Iranians that something significant was happening. Additionally, Stuxnet disabled the centrifuges' safety system, which was designed to turn off centrifuges before they would be damaged.

The next version of the malware, Stuxnet 1.001, was designed to target the frequency converters rather than only the valves, as Stuxnet 0.5 did. Targeting the frequency converters was a more direct and faster attack method. Like its predecessor, this version of the malware started with a reconnaissance phase of 13 days before it launched the attack. First, the frequency of the converters was increased to 1,410 Hz. After fifteen minutes, the malware returned the frequency to 1,046 Hz, which is the converters' normal operating frequency. Over the next 26 consecutive days, the malware did not change the frequency but collected data on the converters' setup. A second attack followed, during which the malware decreased the frequency to 2 Hz. After fifteen minutes, the malware would return the frequency to 1,046 Hz (i.e., the normal operating frequency). The attack sequence repeated itself every 26 days.

Both versions of Stuxnet collected critical data from a centrifuge's normal operation. The malware would later use this data to project to the Iranian technicians operating the installation. This tactic is typically referred to as a "man-in-the-middle attack." By doing so, the Iranian technicians saw normal operating speeds on their displays, while in reality, the centrifuges were manipulated and spun out of control. By using this tactic, Stuxnet was able to remain undetected. The attacks did not occur unwittingly for the Iranians, as the amount of uranium produced at Natanz decreased. Due to the deceptive character of Stuxnet, however, the technicians could not identify the cause for the decrease in production.

Detection of Stuxnet

On June 17, 2010, Ulasen first detected an unknown malware on a computer belonging to one of its customers in Iran.[55] His initial analyses of the malware allowed him to make several important observations regarding its configuration and mode of operation.[56] First, the malware, which was then called Rootkit.TmpHider and SScope.Rookit.TmpHider.2 exploited the Windows OS by installing two drivers (i.e., mrxnet.sys and mrxcls.sys) to inject code into the system processes and hide the malware itself. Second, both drivers used genuine digital certificates that were digitally signed by Realtek Semiconductor Corp, allowing the malware to bypass automated detection programs. Third, the analyses indicated that the malware used a USB flash drive for propagation.[57] These significant findings persuaded Ulasen to contact Microsoft and Realtek to inform them of the threat he had discovered.[58]

However, when both companies did not respond to Ulasen's calls, he published his findings on VirusBlokAda's website and the Wilders Security Forums.[59] By posting his findings on the forum, Ulasen alerted some of the world's leading cybersecurity experts and influenced them to reverse engineer this unknown malware and explain its purpose.

Symantec made one of the most significant efforts in that regard and dedicated a highly experienced research team to the task. The team reverse-engineered the malware over three months.[60] In their report, the Symantec team renamed the malware "W32.Stuxnet," which comes from splicing keywords found in the malware (.stub and mrxnet.sys).[61] One of the key findings of the Symantec team was that Stuxnet was designed to target Siemens SIMATIC WinCC SCADA systems.[62] A second finding pointed more toward Iran, as the malware only activated itself if an installation had a minimum of 33 frequency converters, which had to be made by the Iranian company Fararo Paya or Finish Vacon.[63] A third finding indicated that besides targeting only frequency converters from those two companies, Stuxnet only targeted frequency converters that ran between 807 and 1201 Hz. The US Nuclear Regulatory Commission has regulated the export of frequency converters that run above 600 Hz because they can be used to enrich uranium.[64] Connecting these three findings allowed the Symantec team to suggest that Stuxnet was developed to target Iran's uranium enrichment facility at Natanz.[65]

The discovery of Stuxnet on June 17, 2010, provides little guidance on the timeline of the operation. Essential questions remained: How long did the malware operate undetected, when did the operation start, and to whom was the operation attributable? To try to answer these questions, the Symantec team reverse-engineered the malware and found that the inclusion of two authentic digital certificates and four zero-day exploits, as well as the malware's complex behavior, suggested that the development of the code would have taken approximately 10,000 working hours. Moreover, the substantial resources required to plan, prepare, and execute the operation, coupled with the sensitive nature of the target, point toward the involvement of one or more nation-state actors, allegedly Israel and the US.

Analysis

Policy

The Iranian nuclear program poses a significant threat to Israel, the US, and global security. In that regard, ensuring that Iran will never acquire a nuclear weapon is of vital interest to Israel and the US[66] The 2006 US National Security Strategy (NSS) noted, "Iran has violated its Non-Proliferation Treaty (NPT) safeguards obligations and refuses to provide objective guarantees that its nuclear program is solely for peaceful purposes."[67] The US government, therefore, "will continue to take all necessary measures to protect our national and economic security against the adverse effects of their bad conduct."[68] With the uranium enrichment facility at Natanz, the Iranian regime not only violated the NPT but also did not respond to calls from the UNSC to suspend the development of its nuclear program without delay.[69] Subsequent economic sanctions imposed by the US also turned out to be ineffective.[70] As a result, Israel tried to persuade the US to assist in conducting an air strike on the site at Natanz. The US declined the request, cautious not to repeat the same mistakes made in the Iraqi case in 2003.[71] President Bush, however, offered the Israelis an alternative option aimed at delaying uranium enrichment rather than destroying the site itself.

Military strategy

Drafting response options to cope effectively with Iran's nuclear program was difficult because it required coercive action (i.e., stop doing what you are doing) rather than deterrence (i.e., do not do something). Of the three options available – covert sabotage on Iranian soil, an airstrike, and a cyberattack – President Bush selected the cyberattack. This new attack method fell between coercive diplomacy and overt military force. While the former had proven unsatisfactory, the latter would have profound implications. Thus, the cyberattack served as an expansion of strategic choice for the American president.

In short, the attack would first generate effects in the virtual dimension by intruding on the LAN and manipulating the PLCs. The malware would then generate modest effects in the physical dimension by damaging the centrifuges slowly and over an extensive period of time, all without the risk of any human casualties or collateral damage and with minimal chances of detection and attribution. In other words, the attackers had chosen a protracted and clandestine execution phase in which the inflicted damage to the centrifuges would be relatively small, as opposed to a short and more destructive attack executed covertly and followed by coercive diplomacy.

Suitability: could the attackers execute the operation with the available resources?

The attackers successfully conducted the operation with the available resources. First, the attackers had to collect the required information about the Natanz site. Through various intelligence collection tools, they got the required information on the site's components, configuration, and mode of operation. With that information available, the attackers could build malware that could exploit several zero-day vulnerabilities in the site's network and operate fully autonomously, as commanding and controlling the malware was impossible. Additionally, with the materials retrieved from Libya, the attackers could build a replica of the Natanz site, which allowed them to conduct full-dressed rehearsals. The combination of these actions provided the attackers with an information advantage over the Iranians, as they now had a virtual weapon that was thoroughly tested and ready for use.

The attackers had to ensure that the malware could infiltrate the Natanz site undetected. Because the site's LAN was air-gapped, the attackers had to physically deliver the malware to the site and inject it into one of its computers. The attackers used Erik van Sabben, the Dutch agent whom the AIVD had recruited, to accomplish this.

A second crucial component of the operation was preventing the Iranians from becoming suspicious once they noticed that the centrifuges were being damaged. To this end, the attackers ensured that Stuxnet's mode of operation was selective in its target selection, as it only targeted a certain number of centrifuges in one attack wave. Furthermore, each centrifuge was damaged slowly but surely. Stuxnet also collected critical data from a centrifuge's normal operation, which the malware would later use to project normal system function to the Iranian technicians operating the installation. In doing so, Stuxnet was able to remain undetected while inflicting significant damage to the site's centrifuges.

Feasibility: did achieving the objectives create the strategic effects contributing to the desired end?

The purpose of damaging the centrifuges was to decrease uranium enrichment, thereby delaying Iran's nuclear weapons program. To this end, the attackers conducted several tactical activities that led to the achievement of several collateral effects. First, attacking computers that ran Siemens SIMATEC Step 7 .DDL led to the virtual effect of manipulating the PLCs. This, in turn, resulted in the physical effect of the centrifuges spinning out of control. Furthermore, a typical cognitive effect was generated by fooling the Iranian technicians by projecting normal system function (i.e., a spoofing attack). All three effects should be considered tactical effects. Second, the manipulated and thus dysfunctioning centrifuges resulted in a significant decrease in uranium production by the infected centrifuges. For example, Stuxnet reportedly destroyed 1,000 centrifuges, which is a significant physical effect, particularly compared to other less attractive options presented to Bush. However, decreasing the total amount of enriched uranium was not achieved. In fact, the production of enriched uranium increased during the period Stuxnet was employed because the Iranians could improve the efficiency of the remainder of the centrifuges (i.e., 8,000 in total) that Stuxnet did not infect.[72] Granted, the facility probably would have produced much more uranium had the operation not been conducted, but it remains unclear whether this second-order effect, which was the strategic effect, led to the achievement of the strategic objective of delaying the Iranian nuclear arms program.

Acceptability: did the strategic effects sought to justify the objectives pursued and consider tangible and intangible costs?

To determine whether the objectives meet the conditions of acceptability based on consequences for the attackers, it is crucial to look at three distinct forms of costs, namely the costs of a potential nuclear incident, costs in case the malware would spread, and infect computers outside the Natanz facility (i.e., collateral damage), and the costs for the attackers of losing their favorable balance of power in the region.

The virtual attack method was not without risk. First, the attackers had to be sure that manipulating the centrifuges would not lead to a nuclear incident. Analysis during the preparation of the attack had indicated that this would not be the case.

Second, once a computer containing Stuxnet would be connected to the Internet, the malware would spread around the web in search of other target systems. This would affect the computers of other organizations or even private citizens, potentially causing collateral damage. More importantly, it would alert governments or private companies who could expose the operation and cause it to lose its clandestine character. Furthermore, if the Stuxnet code were reverse-engineered, specific features in the code, such as timestamps or notes that the attackers had forgotten to delete, could lead to attribution and potentially lead to retaliation. Finally, the employment of such a highly sophisticated virtual attack capability carried the risk of exposing the attackers' sophisticated cyber capabilities to a global audience, as described by a former CIA officer:

> Once it's out there, it's like using your stealth fighter for the first time-you've rung that bell and you can't pretend that the stealth fighter doesn't exist anymore. So, the question is, which air battle do you really want to use that stealth fighter for?[73]

The malware's discriminate behavior was designed to minimize these risks by exploiting specific Iranian vulnerabilities. For example, Stuxnet only targeted computers that ran Siemens SIMATEC Step 7 .DDL that were used in Iranian nuclear facilities employing the IR-1 centrifuges. By doing so, the attackers significantly reduced the number of computers that could get infected if the malware accidentally spread to the Internet.

To maintain the attackers' favorable balance of power, the operation had to remain undetected for as long as possible. Once the operation was detected, the attackers wanted to prevent attribution at all costs. To this end, the attackers applied various measures of operational security that prevented the Iranians from detecting or attributing the operation. During the planning and preparation phases of the operation, the attackers extensively applied the principle of security. For example, the attackers conducted the full-dressed rehearsals at Oak Ridge, a remote location that was highly secured and inaccessible to unauthorized personnel. This location allowed the attackers to extensively test the malware's functions, enabling them to make the necessary adjustments to the codes they used before their employment. The full-dressed rehearsals allowed to determine how the frequency converters could be manipulated without making the Iranian engineers suspicious. This knowledge contributed to the operation's covert character, as detecting the attack became extremely difficult. During the attack phase, an element of deception was added to the operation by projecting normal operating data to the Iranian engineers that made them believe the centrifuges were functioning normally.

The operation was exposed when an employee of an Iranian company associated with the Natanz facility connected an infected computer to the Internet. Since the malware did not recognize the changed environment, it spread and infected approximately 200,000 computers. This alarmed several leading IT security firms, including Symantec. One of their analytical teams successfully re-engineered the malware and concluded that it targeted a particular type of PLC the Iranians use in their nuclear facilities. With the operation exposed, the analytical team looked into possible attribution. Several features found in the Stuxnet code indicated a possible link with Israel. Still, they were insufficient to officially attribute the operation to Israel, as these features could have been false flags and, thus, an act of deception. Other details in the code pointed toward the US For example, the malware had an expiration date of January 2009, corresponding with the inauguration of Barack Obama as President of the US This could indicate that a new president had to be briefed on the operation and decide whether to continue. Besides these features, no additional evidence has been found to make attribution to Israel or the US possible, and neither did these countries confirm their involvement.

Even though neither the US government nor the Israeli admitted their involvement, several US government officials have anonymously claimed that these two countries were responsible for the operation.[74] More importantly, these officials claimed that Olympic Games was part of a much bigger US operation, code-named Nitro Zeus. This operation was designed to allow the NSA persistent access to virtually all of Iran's critical infrastructure, including the electricity net, health care, and transportation sector. In case of war with Iran, the NSA would be able to paralyze all these systems in support of US political objectives. The broad scope of Nitro Zeus and the risks of these sophisticated cyber capabilities persuaded NSA officers to speak about the mission publicly, thereby further exposing US covert operations and capabilities.[75]

In sum, whether the strategy's inherent logic proved valid is doubtful. The ends, ways, and means answer the questions of suitability and feasibility, but not acceptability. Stuxnet was designed in such a way as to avoid detection while simultaneously inflicting its destructive capability on Natanz's centrifuges over an extensive period. The cyberattack itself made a physical attack on Iranian soil superfluous, while such an attack was seen, from the beginning, as highly unattractive by the US Although the operation remained undetected for an extensive period, during which an estimated 1,000 centrifuges were damaged, the attackers could not decrease the total amount of enriched uranium. Therefore, it's questionable whether the strategic effect of delaying the Iranian nuclear program was achieved. From a different perspective, it remains unclear to what extent the operation's goal, i.e., the reduction of Iran's nuclear threat, was attained, as the operation's outcome was, at best, modest. Finally, the public exposure in the aftermath of the operation indicates that the attackers' own personnel involved in the operation question the acceptability of the employment of these highly sophisticated and potentially lethal cyber capabilities, particularly when considering that these may be used against the US and its allies.

Risk

What follows is an assessment of the risks associated with the operation. This assessment analyzes the consequences of success and failure; examines the ends, ways, and means relative to the strategic environment; and determines which factors impacted the generation of strategic effects.

Which assumptions did the attackers make in their strategy, and what was the effect of any wrong assumption?

The attackers rightly assumed that adding more spreading mechanisms to the malware would significantly increase the risk of exposure. To prevent the risk of exposure, Stuxnet was designed to minimize the risk of collateral damage. A questionable aspect in this regard is whether the attackers became increasingly willing to accept the risk of exposure in case the malware would be exposed to the Internet. Over time, the attackers made several modifications to the Stuxnet malware. These modifications made it more aggressive. For example, a 2010 version of Stuxnet included more spreading mechanisms than a 2009 version. More spreading mechanisms might have been added to increase the chance of reaching their target more quickly. The downside of these aggressive spreading mechanisms was that Stuxnet would target more indiscriminately, which increased the chances of unintentionally infecting other machines. Indeed, one such spreading mechanism was the. LNK exploit that ultimately led to Stuxnet's discovery. Given the significant skill displayed by the creators of Stuxnet, they most likely would have been aware of the increased risk they were taking but were willing to take. Indeed, the Israelis were reportedly unsatisfied with the results produced by Stuxnet 0.5 and decided to launch version 1.001 without coordinating this with the NSA. This new version was sent to five Iranian companies associated with Natanz. One of these companies became infected with Stuxnet and unknowingly connected one of its devices to the Internet. In other words, the malware did not spread from Natanz to the outside world but rather from another company used by the attackers to increase the chances of infecting the installation.

Did the attackers incorporate flexibility or adaptability into their strategy? If required, how could they have modified their strategy, and at what cost?

Considering the complexity of the operation, the limited risk involved to attackers, and the fact that they could successfully execute the operation over an extensive period without being detected, it is doubtful that any modifications could have improved the operations' outcome.

How did other actors react to the operation?

Once Stuxnet was unleashed on the Internet, several leading computer security firms reverse-engineered the malware and published reports with their findings. Additionally, several news agencies started to publish articles on the operation. For example, *The Washington Post* published an article in 2012 in which anonymous US government officials claimed that the US was behind the attack.[76] The same sources claimed that Israel was involved as well.[77] Iran also responded to the operation's exposure and stated that its nuclear program was designed for peaceful purposes, including generating electricity.[78] Kazem Seddiqi, a senior Iranian cleric, argued that "despite all plots and mischievous behavior of the Western countries … Iran did not withdraw one iota from its rights."[79] Additionally, in 2012, the Iranian hacker group "Shamoon" conducted several retaliatory cyberattacks on financial institutions in the US, including Bank of America, J.P. Morgan/Chase, and Wells Fargo, as well as the Saudi Arabian Oil Company (Saudi Aramco).

In addition to the media attention focused on attribution, a separate debate started among cyber security officials about the implications of the operation. For example, Rafal Rohozinski, a cyber security researcher, asserted that the launch of Stuxnet "officially signals the beginning of the cyber arms race in practice and not in theory."[80] Other scholars and practitioners argued that operations such as Olympic Games would significantly impact domestic and international law, Internet governance, as well as international relations.[81] Indeed, the operation had created a precedent and signaled that employing a cyberattack was something an actor could do if there were problems with another.

However, the trending topic in these discussions was that the operation had demonstrated what is possible beyond imagination. Using computer code as a virtual attack capability represents a clear example in which the attackers caught the Iranians off-guard by exploiting a zero-day Windows OS vulnerability. It is doubtful whether the Iranians could have done anything to prevent the exploitation of that Windows OS vulnerability, as it was unknown to the public, including Microsoft itself. In other words, the Iranians were completely unaware of the security gap in their systems and would have likely been counting on the effectiveness of Windows security patching. Therefore, the successful exploitation of the zero-day vulnerability in the Windows OS took the Iranians entirely by surprise and allowed the attackers to manipulate the centrifuges for an extensive period.

How did unforeseen factors play a role in the strategy?

The operation was mostly executed according to the plan. Indeed, the reconnaissance and delivery phases were completed without any problems. The same applies to the malware's lateral movement in search of the Siemens SIMATEC Step 7 .DDL, as well as the deception of the site's engineers by projecting operating data that made them believe the centrifuges operated as programmed. Even the eventual exposure of the malware seems to have been a risk willingly taken, at least from an Israeli perspective, given their significant programming skills that must have made them aware of the risks involved in adding additional spreading mechanisms to the malware. Thus, no indications of chance or friction that played a role in this strategy were found.

Tactics

Did the attackers achieve relative superiority throughout the operation?

Operation Olympic Games is an excellent case study to extrapolate relative superiority theory, as the employment of a virtual attack capability is not about Clausewitzian troop numbers. Relative superiority theory was developed explicitly for Direct Action (DA) and was applied to historical case studies characterized by special operations conducted by air, land, or sea. Moreover, for the special operations selected for his work, Admiral (Ret.) William McRaven defined mission success in terms of effects generated in the physical dimension. Operation Olympic Games is very different, which requires that the potential ways of achieving relative superiority must be aligned with the ability to conduct special operations in cyberspace, a domain of military operations that did not exist when the theory was developed.

The attackers achieved relative superiority by having an information advantage coupled with the employment of a virtual attack capability that operated undetected. The protracted intelligence collection phase of the operation allowed the attackers to identify the site's unique vulnerabilities, develop the required malware to exploit these vulnerabilities, test its mode of operation, and deliver it to the Natanz site. By doing so, the attackers could take the Iranians by surprise and start with the protracted destructive phase while undetected. Indeed, the employment of Stuxnet illustrates how effects were generated in the virtual and, subsequently, physical and cognitive dimensions. The fact that there was no general understanding of even the existence of such highly sophisticated attack capabilities illustrates the uniqueness of this mission. Also, by employing a virtual capability, the attackers reduced the risk to the force to just one individual, i.e., van Sabben, who infiltrated the site.

The attackers could operate for such a long time by significantly decreasing the area of vulnerability from the onset of the operation. The ability to do so is inherent in the character of the Stuxnet malware, which allowed for the infiltration of the site's LAN and infected the Siemens SIMATEC Step 7 .DDL. Thus, the ultimate control of the centrifuges was obtained undetected. By employing these actions at the onset of the operation, the attackers achieved relative superiority. Having achieved this condition at this early stage in the operation meant that the chance of mission success outweighed the chance of mission failure. Figure 5.2 illustrates

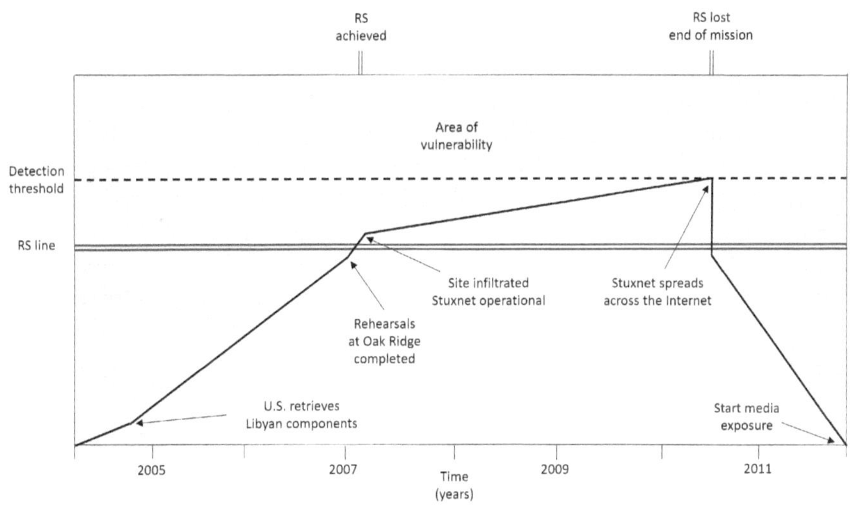

Figure 5.2 Relative superiority graph for the cyberattack at Natanz.

this by the significant steep curve that reaches far above the Z-axis (i.e., relative superiority line), thereby decreasing the area of vulnerability substantially.

If achieved, how did the attackers sustain their relative superiority throughout the operation? If not, why did they fail?

The attackers sustained their relative superiority by conducting a man-in-the-middle attack and because of the malware's discriminatory targeting. After the effective intrusion of the LAN and the subsequent takeover of the Siemens SIMATEC Step 7 .DDL, Stuxnet started to inflict damage on the centrifuges by employing its destructive capability. The operation, however, remained vulnerable to detection, as the disruptive actions could alert the Iranian engineers who operated the installation. Another vulnerability was the risk of exposing the operation if Stuxnet was unleashed online if an infected computer were connected to the Internet. Meanwhile, to complete the mission, relative superiority had to be sustained until the desired level of damage to the site was inflicted. Two specific features were added to the malware, allowing it to achieve that. First, Stuxnet was able to project manipulated data about the centrifuge's mode of operation to the Iranian technicians operating the installation, which made them believe the centrifuges operated at normal operating speed. Second, Stuxnet employed discriminate targeting, meaning only computers running Siemens SIMATEC Step 7 .DDL would get infected. Thus, the attackers achieved relative superiority because they had collected reliable intelligence, maintained excellent operational security, conducted extensive tests of the malware in a controlled environment, selected their best personnel, and, above all, employed superior technology. These factors combined allowed the attackers to completely circumvent the friction of war, as a virtual weapon could enter an Iranian nuclear facility undetected, causing substantial damage to the centrifuges.

Once Stuxnet was unleashed on the Internet, the Iranians quickly learned about the malware and shut the installation down. This meant that detection immediately led to the loss of relative superiority. Importantly, detection meant the end of the mission, as the site's systems received the required security patches shortly after that. Figure 5.2 illustrates this by the significant declining curve reaching far below the Z-axis, thereby making the area of vulnerability very large. For the attackers, the area of vulnerability remained large for an extensive period, even after the mission had ended, as they wanted to prevent attribution. In 2012, however, the *Washington Post* published an article indicating that the US and Israel were behind the operation. In 2016, the documentary Zero Days was released, in which several alleged anonymous NSA officers claimed US and Israeli involvement. Consequently, Operation Olympic Games illustrates the difficulty of maintaining the desired level of secrecy, even after the operation had been completed. In this case, it was not because of the trails that the attackers left behind but because of their own personnel (i.e., the NSA officers who spoke about the mission publicly), who became overly concerned with the covert activities they were involved in.

Conclusion

Operation Olympic Games illustrates how cyberspace allowed for a different approach to special operations. The design of the operation was highly unique, and the attackers employed superior tactical action. In that regard, a high level of strategic sense was shown, as demonstrated by converting tactical performance into strategic effects using cyberspace. However, although 1,000 centrifuges were successfully destroyed (first-order effect), it remains uncertain whether the second-order effect, delaying uranium enrichment, was achieved. Therefore, the operation cannot be used as an example in which the attackers fulfilled the strategic promise of special operations through the careful alignment of policy, strategy, and tactics.

The operation also sheds new light on the ancient principle of economy of force. The employment of Stuxnet as a virtual attack capability demonstrates the ability to conduct special operations with an extremely limited number of troops with "boots on the ground." In fact, in this "first" case, just one man! In that regard, the operation illustrates that special operations conducted through cyberspace have the potential to contribute significantly to achieving strategic objectives disproportionate to the forces that must be physically deployed into a hostile environment. In other words, the emergence of cyberspace will allow policymakers and military commanders to employ special operations with significantly less risk to force than compared to traditional sabotage operations. This highlight possible changes in modern warfare and, thus, in special operations, as the emergence of cyberspace allows policymakers to employ virtual military means as an alternative to classic kinetic action to achieve political objectives.

Notes

1 Council on Foreign Relations, "Iran's Nuclear Program."
2 Ibid.
3 Ibid.
4 United Nations Office for Disarmament Affairs, "Treaty on the Non-Proliferation of Nuclear Weapons."
5 International Atomic Energy Agency, "The Text of the Agreement Between Iran and the Agency for the Application of Safeguards in Connection with the Treaty on the Non-proliferation of Nuclear Weapons."
6 Council on Foreign Relations, "Iran's Nuclear Program."
7 Director of Central Intelligence, "Special National Intelligence Estimate: Prospects for Further Proliferation of Nuclear Weapons," August 23, 1974.
8 Ibid., 38.
9 Kelsey Davenport, "Timeline of Nuclear Diplomacy with Iran, 1967-2023," Arms Control Association, January 2023.
10 Mohammad Javad Zarif, "Tackling the Iran-US crisis: The need for a paradigm shift," *Journal of International Affairs* 60, no. 2 (2007): 73–94.
11 Leonard S. Spector, "Iranian Nuclear Program Remains Major Threat Despite Partial Freeze of Weapons-Relevant Activities Described in New US National Intelligence Estimate," *CNS Feature Story*, December 6, 2007.
12 Congressional Research Service, "Iran's Nuclear Program: Status," *CRS Report*, Updated December 20, 2019.

13 George W. Bush, "The President's State of the Union Address," January 29, 2002.
14 National Intelligence Strategy of the United States of America (2009).
15 Kelsey Davenport, "UN Security Council Resolutions on Iran," Arms Control Association, January 2022.
16 Michael V. Hayden, *Playing to the Edge: American Intelligence in the Age of Terror* (New York: Penguin Book, 2016), 340.
17 Mariusz Antoni Kamiński, "Operation "Olympic Games." Cyber-Sabotage as a Tool of American Intelligence Aimed at Counteracting the Development of Iran's Nuclear Programme," *Security & Defence Quarterly* 2, no. 29 (2020): 65; After President Bush informed PM Olmert that the US would not conduct a strike against the *Al-Kibar* site, the Israelis decided to destroy it themselves by launching Operation Orchard, which is described in Chapter 4 of this book.
18 George W. Bush, *Decision Points* (New York: Crown Publishing Group, 2011), 488.
19 Ibid., 488–491.
20 Ibid., 488–491.
21 David E. Sanger, "US Rejected Aid for Israeli Raid on Iranian Nuclear Site', *New York Times*, January 10, 2009.
22 Kamiński, Operation "Olympic Games," 66.
23 Ibid.
24 William J. Broad and David E. Sanger, "In Nuclear Net's Undoing, a Web of Shadowy Deals," *New York Times*, April 24, 2008.
25 Alireza Jafarzadeh, "Iran and Nuclear Weapons," *C-SPAN*, August 14, 2002.
26 Kim Zetter, *Countdown to Zero Day: Stuxnet and the Launch of the World's First Digital Weapon* (New York: Crown Publishers, 2014).
27 George Tenet speaking at Georgetown University on February 5, 2004.
28 US Department of State, "Designation of National Council of Resistance and National Council of Resistance of Iran under Executive Order 13224," *Press Statement*, August 15, 2003.
29 David Albright, Paul Brannan, Andrea Stricker, Christina Walrond, and Houston Wood, *Preventing Iran from Getting Nuclear Weapons: Constraining Its future Future Nuclear Options* (New York: The Institute for Science and International Security, 2012), 12–13.
30 Joby Warrick, "US Is Said to Expand Covert Operations in Iran," *Washington Post*, June 30, 2008.
31 Zetter, *Countdown to Zero Day.*
32 Nicolas Falliere, Liam O Murchu, and Eric Chien, "W32.Stuxnet Dossier," *Symantec Security Response*, Version 1.3 (November 2010), 1–64.
33 Falliere, O Murchu, and Chien, "W32.Stuxnet Dossier"; Ralph Langner, "417 Data Structures = Cascade Structure = Reported Damage," *Langner Communications GmbH Blog*, December 29, 2010.
34 David Albright quoted in: Zetter, *Countdown to Zero Day*, 311–313.
35 Zetter, *Countdown to Zero Day.*
36 Nicole Perlroth, *This is How They Tell Me the World Ends* (New York: Bloomsbury Publishing, 2020).
37 Ilias Chantzos, Director of Government Relations at Symantec cited in: Thomas Chen, "Stuxnet, the Real Start of Cyber Warfare? Editor's Note," *IEEE Network* 24, no. 6 (2010), 3.
38 Zetter, *Countdown to Zero Day.*
39 David Sanger and Thom Sanker, "N.S.A. Devises Radio Pathway into Computers," *New York Times*, January 14, 2014.
40 Huib Modderkolk, "Sabotage in Iran: een missie in het duisternis," *De Volkskrant*, January 8, 2024.
41 Modderkolk, "Sabotage in Iran."

42 Ibid.
43 Ibid.
44 Zetter, *Countdown to Zero Day*.
45 Kelsey Davenport, "Chronology of Libya's Disarmament and Relations with the United States," Arms Control Association, March 2021.
46 Falliere, O Murchu, and Chien, "W32.Stuxnet Dossier."
47 Ibid.
48 Sergey Ulasen, post on Wilders Security Forums.
49 Modderkolk, "Sabotage in Iran."
50 Timo Steffens, *Attribution of Advanced Persistent Threats: How to Identify the Actors of Cyber Espionage* (Berlin: Springer-Verlag, 2020).
51 Falliere, O Murchu, and Chien, "W32.Stuxnet Dossier."
52 Ibid.
53 Ibid., 3.
54 Falliere, O Murchu, and Chien, "W32.Stuxnet Dossier."
55 Sergey Ulasen, cited in: Eugene Kaspersky, "The Man Who Found Stuxnet – Sergey Ulasen in the Spotlight," November 2, 2011.
56 Sergey Ulasen, post on Wilders Security Forums, July 12, 2010.
57 Ibid.
58 Ulasen cited in: Kaspersky.
59 Ibid.; Sergey Ulasen, post on Wilders Security Forums.
60 Sharon Weinberger, "Computer Security: Is This the Start of cyberwarfare?" *Nature* 474, no. 143 (2011), 142.
61 Falliere, O Murchu, and Chien, "W32.Stuxnet Dossier."
62 Ibid.
63 Zetter, "Clues Suggest."
64 United States Nuclear Regulatory Commission, "Documents Collection, Appendix B to Part 110—Illustrative List of Gas Centrifuge Enrichment Plant Components Under NRC's Export Licensing Authority," November 16, 2022.
65 Liam O Murchu cited in: Zetter, "Clues Suggest."
66 George W. Bush, *National Security Strategy* (Washington, DC: The White House, March 16, 2006).
67 Ibid., 20.
68 Ibid, 20.
69 National Intelligence Strategy of the United States of America (2009).
70 Davenport, "UN Security Council Resolutions on Iran."
71 Bush, *Decision Points*, 488–491.
72 David Albright, Andrea Stricker, and Christina Walrond, "IAEA Iran Safeguards Report: Shutdown of Enrichment at Natanz Result of Stuxnet Virus? Downgrade in Role of Fordow Enrichment Site; LEU Production May Have Decreased," *Institute for Science and International Security*, November 23, 2010.
73 CIA operative cited in: Zetter, *Countdown to Zero Day*, 203.
74 NSA officer cited in: Alex Gibney, "Zero Days," *Magnolia Pictures*, July 8, 2016.
75 Ibid.
76 Ellen Nakashima and Joby Warrick, "Stuxnet was Work of US and Israeli Experts, Officials Say," *The Washington Post*, June 2, 2012.
77 Nakashima and Warrick, "Stuxnet was work of US and Israel."
78 Ibid.
79 Ibid.
80 Rafal Rohozinski cited in: Nakashima and Warrick, "Stuxnet was work of US and Israel."
81 Adam Segal, *Cyber Conflict After Stuxnet: Essays from the Other Bank of the Rubicon* (Vienna: Cyber Conflict Studies Association, 2016).

Bibliography

Albright, David, Paul Brannan, Andrea Stricker, Christina Walrond, and Houston Wood. *Preventing Iran from Getting Nuclear Weapons: Constraining Its Future Nuclear Options* (New York: The Institute for Science and International Security, 2012).

Albright, David, Andrea Stricker, and Christina Walrond. "IAEA Iran Safeguards Report: Shutdown of Enrichment at Natanz Result of Stuxnet Virus? Downgrade in Role of Fordow Enrichment Site; LEU Production May Have Decreased," Institute for Science and International Security, November 23, 2010.

Broad, W. J., and D. E Sanger. "In Nuclear Net's Undoing, a Web of Shadowy Deals," *New York Times*, April 24, 2008, https://www.nytimes.com/2008/08/25/world/25nuke.html.

Bush, George W. "The President's State of the Union Address," January 29, 2002, https://georgewbush-whitehouse.archives.gov/news/releases/2002/01/20020129-11.html.

Bush, George W. National Security Strategy (Washington, DC: The White House, March 16, 2006), https://www.comw.org/qdr/fulltext/nss2006.pdf.

Bush, George W. *Decision Points* (New York: Crown Publishing Group, 2011).

Chen, Thomas. "Stuxnet, the Real Start of Cyber Warfare? Editor's Note," *IEEE Network* 24 (2010): 6.

Congressional Research Service. "Iran's Nuclear Program: Status," CRS Report, Updated December 20, 2019, https://crsreports.congress.gov/product/details?prodcode=RL34544.

Council on Foreign Relations. "Iran's Nuclear Program," March 10, 2010, https://web.archive.org/web/20120220182315/http://www.cfr.org/iran/irans-nuclear-program/p16811.

Davenport, Kelsey. "Chronology of Libya's Disarmament and Relations with the United States," Arms Control Association, March 2021, https://www.armscontrol.org/factsheets/LibyaChronology.

Davenport, Kelsey. "UN Security Council Resolutions on Iran," Arms Control Association, January 2022, https://www.armscontrol.org/factsheets/Security-Council-Resolutions-on-Iran.

Davenport, Kelsey. "Timeline of Nuclear Diplomacy with Iran, 1967-2023," Arms Control Association, January 2023, https://www.armscontrol.org/factsheets/Timeline-of-Nuclear-Diplomacy-With-Iran.

Director of Central Intelligence. "Special National Intelligence Estimate: Prospects for Further Proliferation of Nuclear Weapons," 23 August 1974, https://web.archive.org/web/20120226164727/http://www.gwu.edu/~nsarchiv/NSAEBB/NSAEBB240/snie.pdf.

Gibney, Alex. "Zero Days," *Magnolia Pictures*, July 8, 2016.

Hayden, Michael. *Playing to the Edge: American Intelligence in the Age of Terror* (New York: Penguin Book, 2016).

International Atomic Energy Agency. "The Text of the Agreement Between Iran and the Agency for the Application of Safeguards in Connection with the Treaty on the Non-proliferation of Nuclear Weapons," December 13, 1974, https://www.iaea.org/publications/documents/infcircs/text-agreement-between-iran-and-agency-application-safeguards-connection-treaty-non-proliferation-nuclear-weapons.

Jafarzadeh, Alireza. "Iran and Nuclear Weapons," C-SPAN, August 14, 2002, https://www.c-span.org/video/?172005-1/iran-nuclear-weapons.

Kamiński, Mariusz Antoni. "Operation 'Olympic Games' Cyber-Sabotage as a Tool of American Intelligence Aimed at Counteracting the Development of Iran's Nuclear Programme," *Security & Defence Quarterly* 2 (2020): 29.

Langner, Ralph. "417 Data Structures = Cascade Structure = Reported Damage," Langner Communications GmbH Blog, December 29, 2010, http://www.langner.com/en/2010/12/29/417-data-structures-cascade-structure-reported-damage/.

Modderkolk, Huib. "Sabotage in Iran: Een Missie in het Duisternis," *De Volkskrant*, January 8, 2024, https://www.volkskrant.nl/kijkverder/v/2024/sabotage-in-iran-een-missie-in-duisternis~v989743/?referrer=https%3A%2F%2Fwww.google.nl%2F.

Nakashima, Ellen, and Joby Warrick. "Stuxnet was Work of US and Israeli Experts, Officials Say," *The Washington Post*, June 2, 2012, https://www.washingtonpost.com/world/national-security/stuxnet-was-work-of-us-and-israeli-experts-officials-say/2012/06/01/gJQAlnEy6U_story.html.

National Intelligence Strategy of the United States of America. (2009) Available at: http://fas.org/irp/offdocs/nis2009.pdf.

Perlroth, Nicole. *This Is How They Tell Me the World Ends* (New York: Bloomsbury Publishing, 2020).

Sanger, D. E. "US Rejected Aid for Israeli Raid on Iranian Nuclear Site', *New York Times*, January 10, 2009, https://www.nytimes.com/2009/01/11/washington/11iran.html.

Sanger, David, and Thom Sanker. "N.S.A. Devises Radio Pathway into Computers," *New York Times*, January 14, 2014, https://www.nytimes.com/2014/01/15/us/nsa-effort-pries-open-computers-not-connected-to-internet.html.

Segal, Adam. *Cyber Conflict After Stuxnet: Essays from the Other Bank of the Rubicon* (Vienna: Cyber Conflict Studies Association, 2016).

Spector, Leonard S. "Iranian Nuclear Program Remains Major Threat Despite Partial Freeze of Weapons-Relevant Activities Described in New US National Intelligence Estimate," CNSmFeature Story (December 6, 2007). https://web.archive.org/web/20120815203104/http://cns.miis.edu/stories/071206.htm.

Steffens, Timo. *Attribution of Advanced Persistent Threats: How to Identify the Actors of Cyber Espionage* (Berlin: Springer-Verlag, 2020).

Tenet, George. Speaking at Georgetown University on February 5, 2004. A transcript of his talk is available at: https://www.cia.gov/news-information/speeches-testimony/2004/tenet_georgetownspeech_02052004.html.

Ulasen, Sergey. Post on Wilders Security Forums, July 12, 2010, https://www.wilderssecurity.com/threads/rootkit-tmphider.276994/#post-1711587.

Ulasen, Sergey. Cited in Eugene Kaspersky, "The Man Who Found Stuxnet – Sergey Ulasen in the Spotlight," November 2, 2011, https://eugene.kaspersky.com/2011/11/02/the-man-who-found-stuxnet-sergey-ulasen-in-the-spotlight/.

United Nations Office for Disarmament Affairs. "Treaty on the Non-Proliferation of Nuclear Weapons," 1995, https://treaties.unoda.org/t/npt.

US Nuclear Regulatory Commission. "Documents Collection, Appendix B to Part 110—Illustrative List of Gas Centrifuge Enrichment Plant Components Under NRC's Export Licensing Authority," November 16, 2022, https://www.nrc.gov/reading-rm/doc-collections/cfr/part110/part110-appb.html.

US Department of State. "Designation of National Council of Resistance and National Council of Resistance of Iran under Executive Order 13224," *Press Statement*, August 15, 2003, https://2001-2009.state.gov/r/pa/prs/ps/2003/23311.htm.

Warrick, Joby. "US Is Said to Expand Covert Operations in Iran," *Washington Post*, June 30, 2008, https://www.washingtonpost.com/wp-dyn/content/article/2008/06/29/AR2008062901881_pf.html.

Weinberger, Sharon. "Computer Security: Is This the Start of Cyberwarfare?" *Nature* 474 (2011).

Zarif, Mohammad Javad. "Tackling the Iran-US Crisis: The Need for a Paradigm Shift," *Journal of International Affairs* 60, no. 2 (2007): 73–94.

Zetter, Kim. *Countdown to Zero Day: Stuxnet and the Launch of the world's First Digital Weapon* (New York: Crown Publishers, 2014).

Part III

Case Studies in Indirect Special Operations

6 The Russian Annexation of Crimea

Introduction

The first three case studies described in Chapters 3, 4, and 5 focused on direct special operations. In this chapter, the attention shifts to the second type: indirect special operations. In this case, Russia successfully conducted an operation to destabilize the Ukrainian regime and annex the Crimean Peninsula. What stood out was that the Russians achieved their political objective by using local state and nonstate partners rather than through unilateral action. This concept of operation had specific relevance to the political and security environment in Ukraine, as it provided Russian policymakers with an option between economic coercion and overt military intervention.

Given the importance of the political objective, it was a strategic operation initiated by President Vladimir Putin, that was planned, prepared, and executed in a short period of time, and with limited resources. By doing so, Putin sought to advance Russia's national interests of returning Crimea to Russia through an unconventional use of military force.

Background

Throughout history, old empires contested Ukraine's territory and changed its borders several times.[1] Starting in the eighteenth century, the region was composed of the southern coastal area north of the Black Sea and Donetsk to the east, also known as "*Novorossiya*," which includes the port city of Odesa, and the northern area surrounding Kharkiv. Catherine the Great, then ruling the Russian Empire, took the region from the former Ottoman Empire and colonized it aggressively. By the end of the First World War, Russia and the Central Powers signed the Peace Treaty of Brest-Litovsk, through which Ukraine gained independence.[2] The western part of Ukraine, which had not been part of Russia, was divided among several countries, including Poland. All the territory then became occupied twice: first by the Soviet Union in 1939, followed by the invasion of Nazi Germany in 1941. Finally, after the German defeat in 1945, Ukraine became part of the Soviet Union.

Next to the historical changes in Ukraine's borders, three noteworthy events shaped its relationship with Russia. The first one is the "Great Famine" of

DOI: 10.4324/9781003365112-9

1932–1933, which resulted from Soviet leader Josef Stalin's coercive policy to force Ukrainian farmers to participate in a concept called "collective farming."[3] As a result, approximately four million Ukrainians starved to death.[4] Second, Stalin implemented an aggressive policy to purge, exile, or execute Ukrainian citizens who were in favor of Ukrainian independence. The final tragic event was World War II (WWII). Although a substantial number of Ukrainians initially welcomed the German invaders, their brutality forced the Ukrainians to join the Red Army in its fight against the Nazis. At the end of the war, an estimated eight million Ukrainians had lost their lives, less than 20 percent of whom were combat causalities. The end of WWII left Ukraine in a desperate state, with its infrastructure destroyed and even more human suffering caused by more famine.

In the following years, Ukraine became an important region for the Soviet Union's military-industrial base and key military installations. Although many Ukrainians held influential positions in the political system, economic sector, and defense industry, a distinct Ukrainian identity developed over time. When it became clear that the Soviet Union would no longer continue to exist, many Ukrainians began to claim their sovereignty. As a result, Ukraine held a referendum on December 1, 1991. The outcome could not have been more precise, as 90 percent of the votes favored independence. In addition to the referendum, the Ukrainians elected Leonid Kravchuk as their first president.[5]

After Ukraine became independent from the Soviet Union, the population expected an era of wealth and prosperity. Contrary to these expectations, a significant economic decline hit the country while it struggled to adapt to the new geopolitical situation. Additionally, the country became highly vulnerable to corruption, particularly among the elite. For example, between 1995 and 1999, the global Corruption Perceptions Index (CPI) listed Ukraine as 75 out of 99, indicating a very high level of corruption.[6] As a result, there was a growing discontent among the population. Although Ukraine's economic conditions improved by the end of the twentieth century, public discontent and corruption remained.

In 2002, Ukraine's second president, Leonid Kuchma, appointed Viktor Yanukovych prime minister. In 2004, Yanukovych ran for presidency against Viktor Yushchenko. Initially, the pro-Russian Yanukovych won the elections. Quickly thereafter, reports indicated widespread fraud.[7] As a result, mass street protests and civil unrest, known as the Orange Revolution, broke out in the country.[8] After the *Verkhovna Sud Ukrayiny* (Ukrainian Supreme Court) revoked the outcome of the election, reelection took place in December. [9] During these elections, Yushchenko was poisoned with TCDD dioxin in an unsuccessful attempt to assassinate him.[10] Finally, Yushchenko received most of the votes and was inaugurated in January 2005.

During Yushchenko's presidency, public discontent and corruption remained high. He, therefore, fired his government, including Prime Minister Yulia Tymoshenko, on September 8, 2005, less than a year after its installation.[11] Because his political party was losing popularity, Yushchenko appointed his rival Yanukovych as his new prime minister.[12] The appointment of Yanukovych was also an attempt to improve the relations with Russia. Moving forward, Ukrainian politics was characterized by power play between these three key politicians and heavily influenced by the negotiations and disputes with Russia over oil and gas imports.

The years of political turmoil under President Yushchenko and continuous fights with Tymoshenko allowed Yanukovych to regain power. In the 2010 presidential elections, judged by observers as free and fair, he received most of the votes and became president.[13] His political rival, Tymoshenko, was accused of abusing power and sentenced to seven years in prison.[14] With Yanukovych in office, the possibility of joining the North Atlantic Treaty Organization (NATO) was off the political agenda. Russian President Medvedev viewed this development as positively affecting the improved relationship between the two countries.[15]

Because Yanukovych realized that a potential European Union (EU) membership would significantly strengthen his position and gain public support, he continued negotiations with Brussels. For many former Eastern European countries, joining the EU positively affected their economies.[16] In that regard, Poland served as a powerful example for Ukraine. For instance, between 1990 and 2012, Poland's economy almost doubled in size while Ukraine's economy shrunk by 30 percent.[17] As a result, the GPD per capita in Poland was five times higher than in Ukraine. Thus, for the people of Ukraine, the potential of EU membership equaled prosperity and a better future than they were likely to find in a partnership with Russia.

In 2013, EU expansion included many countries that were once part of the Soviet Union, including the Baltic states. To Ukraine, the EU offered an Association Agreement, which was mainly the result of diplomatic efforts from those former Soviet states. What followed was the EU's Eastern Partnership, which was a number of conferences that included Armenia, Azerbaijan, Belarus, Georgia, Moldova, and Ukraine. According to the EU, the goal of this initiative was to "[building] a common area of democracy, prosperity, stability, and increased interactions and exchanges."[18] Although the EU required the countries involved to make several prominent political and socioeconomic reforms, the prospect of the partnership was to achieve "economic integration in the EU Internal Market, and therefore to the creation of an economic area between the EU and partner countries."[19]

Although the initial cooperation between the EU and Eastern European countries established in 2005 was not a priority for Putin, the statement delivered by the EU during the Warsaw Summit did lead to growing Russian concern.[20] For example, Russia's Foreign Minister Sergei Lavrov stated that, "We are accused of having spheres of influence, but what is the Eastern Partnership, if not an attempt to extend the EU's sphere of influence, including to Belarus?"[21] One of Russia's national interests is maintaining its sphere of influence, often called its "near abroad."[22] Therefore, on November 18, 2011, Russia announced its intention to establish the Eurasian Customs Union (ECU) together with Kazakhstan and Belarus.[23] This Russian initiative, by 2015 renamed Eurasian Economic Union (EEU), was modeled on the EU and would function as a supranational organization focused on integrating economies, customs services, legal systems, and even military capabilities.

For Putin, the ECU was a chance to create a supranational institutional construct from which Russia could exercise its power and influence. Having witnessed the collapse of the Soviet Union and growing discontent in Moscow by late 2011, Putin believed that Russian hegemony had to return.[24] This meant that besides Kazakhstan and Belarus, other former Soviet states, including Armenia, Moldova, and Ukraine, had to remain in Russia's sphere of influence.[25] Shortly thereafter, Russia launched an aggressive, coercive campaign to prevent these countries from falling into the EU's sphere of influence.

Although Russia employed a variety of coercive actions, it mainly targeted Ukraine's energy sector. Such actions were not new, as negotiations over gas distribution go back to when Ukrainian President Kuchma was still in office (1994–2005). Back then, Russia used energy as a weapon to negotiate a favorable outcome in the debate around the size of its Black Sea Fleet based in Sevastopol on the Crimean Peninsula.[26] During the 2005 Orange Revolution, Russia applied the same coercive actions by significantly raising the price of gas.[27] Moreover, Russia warned Ukraine and European countries that Russian gas supplies could be lowered in quantity or even stopped. To make these threats more compelling, Russia dropped the pressure in one of its pipelines, which resulted in reduced gas deliveries to Poland, Austria, Slovakia, and Hungary. In 2009, Russia acted similarly and blocked gas delivery to Ukraine for three consecutive weeks.[28] By doing so, Russia attempted to demonstrate its power and influence on the countries in its sphere of influence and on the West.

In addition to targeting Ukraine's energy sector, Russia inflicted significant economic damage on Ukraine by imposing import restrictions on goods coming from Ukraine. Since almost 25 percent of Ukraine's exports were destined for the Russian market, substantial drops in production and exports led to further economic decline. Russia targeted Ukraine's heavy industry situated in the Donbas region as well as businesses that were pro-EU. As a result, Ukraine's economy arrived at the edge of collapse, as the total foreign debt rose to USD 10.8 billion, which equaled 77 percent of GDP.[29] Yanukovych found himself in a position where he had to choose between two options, neither offering a concrete long-term perspective. The first was to accept the Association Agreement with the EU while realizing that gaining full membership required significant political and economic reforms. The second was to join Russia's ECU, which was still in the conceptual stage. Leading a country that has been characterized by political turmoil, economic decline, and dissent since its independence, neither option provided a clear roadmap to wealth, stability, and prosperity for Ukraine.

On November 21, 2013, Yanukovych abandoned the Association Agreement with the EU and took the path that would lead to increased cooperation with Russia.[30] Part of that cooperation was a Ukrainian–Russian Action Plan that included a significant price reduction for Russian natural gas from USD 400 per 1,000 m^3 to USD 268. Russia also threw Ukraine an economic lifeline by taking over USD 15 billion Eurobonds.[31] Ukrainian Prime Minister Mykola Azarov explained the reason for choosing a partnership with Russia over the Association Agreement with the EU as follows: "The decision to suspend signing the agreement with the EU was difficult but the only possible one in the current economic situation in Ukraine. This decision was dictated solely by economic reasons and is tactical."[32] He added that Ukraine would now focus on renewing its trade relationship with Russia, as it was Ukraine's primary trading partner and supplier of energy.[33]

Was there no alternative to the Ukrainian–Russian Action Plan? Yes, the International Monetary Fund (IMF) could have provided a loan. Such a loan, however, should have been accompanied by structural economic reforms. The IMF had already rejected a USD 15 million loan two years prior because of Ukraine's failure to implement such reforms.[34] According to Azarov, the demands from the IMF included substantial budget cuts as well as several measures that would significantly impact the Ukrainian population through rising energy bills, freezes in salaries, and reductions in energy subsidies. For the Ukrainian government, these conditions were unacceptable; therefore, Ukraine had no choice but to partner with Russia.[35] The desperate situation in which Ukraine found itself in late 2013 was emphasized a few months later by Christine Lagarde, managing director of the IMF, who stated that without the support offered by Russia, "Ukraine was heading nowhere."[36]

Saving Ukraine from bankruptcy was thus not a high priority for the EU. The disappointing outcomes of the Orange Revolution, coupled with the substantial power and influence of Ukrainian oligarchs, high corruption among Ukraine's government officials, and growing skepticism about their willingness to make political and economic reforms, made Brussels reluctant to provide substantial financial support to the country.[37] Importantly, EU officials believed that Yanukovych had surrendered to Putin's coercive actions but would soon find out that the partnership with Russia would not be a sustainable option, as was articulated by one EU official: "Six months down the line, when left alone to deal with Russian pressure, he will come to us on his knees."[38]

Yanukovych's decision to abandon the Association Agreement with the EU and seek closer ties with Russia led to civil unrest among the Ukrainian population. At first, approximately 100,000 people attended a demonstration at the *Maidan Nezalezhnosti* (Independence Square) in central Kyiv, which began on November 21, 2013. These protests signaled the start of the revolution of dignity, popularly known as the Maidan Revolution (initially a hashtag on Twitter). By early December, the number of protesters had grown to around 800,000 people who gathered at the square and occupied Kyiv's city hall. Yanukovych tried to contain the protests by preventing the media from accessing the square and employing the *Berkut* (Ukraine's special police). Putin recognized the threat posed by the protesters and tried to strengthen Yanukovych's position. More importantly, Putin viewed the demonstrations as a Western attempt to overthrow the pro-Russian government with a Western-leaning one. The two men met in Sochi on December 6, when Putin provided Ukraine with a USD 15 billion loan and announced he would reduce the price of Russian natural gas by approximately a third. By doing so, Putin argued, Russia signaled that Ukraine is "without a doubt, in the full sense of the word, our strategic partner and ally."[39]

Rather than serving as a stabilizing factor, the new deal between Putin and Yanukovych led to a further increase in protests in Kyiv. As a result, Yanukovych met again with Putin on January 8, but this time in secret.[40] Subsequently, the *Verkhovna Rada* of Ukraine (Supreme Council of Ukraine) passed restrictive anti-protest laws that led to growing international attention and concern as well as increased demonstrations. Tragically, the situation escalated when protesters stormed regional government offices in western Ukraine. This escalation led to the resignation of Azarov and the annulation of the anti-protest law. At the same time, Yanukovych held talks with the leaders of the opposition and offered them seats in his government. They rejected his offer, and the situation was about to get out of control.

On February 18. 2014, violent clashes between the protesters and Ukrainian police resulted in 26 deaths, ten of which were police officers. Two days later, 44 protesters and three police officers lost their lives at *Maidan Nezalezhnosti*. Videos showed uniformed snipers firing at the protesters.[41] In an ultimate attempt to restore order, the foreign ministers of France, Germany, Poland, and Russia negotiated with Yanukovych and the opposition, successfully striking a deal creating some stability. Russia, however, ultimately refused to sign the agreement.[42] Unsatisfied with the fact that the deal included that Yanukovych would stay in power, the protesters did not step down. When the riot police withdrew from their positions around the presidential palace, Yanukovych fled the country on February 22.[43]

On the same day, protesters took control of the presidential palace and Yanukovych's private residence at *Mezhiyhirya,* which lies just outside Kyiv. Simultaneously, the *Rada* in Kyiv assembled and voted in favor of the official removal of Yanukovych from power and the return to the 2004 constitution. Putin responded by strongly condemning the impeachment of Yanukovych and expressing his support for him. The US and European countries, on the other hand, expressed their support for the Ukrainian people and the democratic decision made by the *Rada*. Finally, former prime minister Tymoshenko was released from prison. In the following days, the *Rada* appointed Olexander Turchynov as interim president, Arseniy Yatsenyuk as prime minister, and announced that new elections were scheduled for May 25, 2014. The *Rada* in Kyiv also voted in favor of banning Russian as the second official language and decided to disband the *Berkut*.

These dynamics made Putin believe that he was losing power and influence over Ukraine and decided that a strong Russian response was vital, as emphasized by Lavrov:

> Would it be acceptable for Russia, considering its international standing, to keep mum and recognize the coup in Ukraine, and to leave Russians and Russian speakers in Ukraine in the lurch after the first order issued by the organizers of the anti-constitutional armed revolt, which was supported by their foreign sponsors, and banned many things that were connected with the Russian language?[44]

By explicitly referring to a "coup," Lavrov delegitimized the democratic Ukrainian government. Furthermore, by emphasizing that it would be unacceptable "to leave

Russians or Russian speakers in the lurch," Lavrov hinted that an urgent response was required.[45] Such a response was soon to come.

The Crimean Peninsula

Starting in the eighteenth century, Crimea was part of the former Ottoman Empire. In 1783, however, Catherine the Great annexed the peninsula and made it part of the former Russian Empire. After the Bolshevik Revolution of 1917, under traditional monarchic leadership, the Russian Empire ended and transformed into the Union of Soviet Socialist Republics (USSR, or Soviet Union), the world's first communist state. Until 1954, Crimea remained part of the Russian Soviet Federative Socialist Republic (SFSR), which was part of the Soviet Union. That year, Soviet leader Nikita Khrushchev wanted to improve the Russian–Ukrainian relationship and issued a decree to transfer Crimea to the Ukrainian Soviet Socialist Republic. Rather than a gesture to improve the relationship, the reason for the transfer of the peninsula was rather administrative and aimed to achieve better economic coordination between the various Soviet republics. The peninsula's population did not receive the transfer well and believed it demonstrated "little consideration of local sentiment or the coherence and legitimacy of the resultant policies."[46] This sentiment shows the historical proximity between Russia and Crimea, characterized by a shared language, culture, and ethnicity. Figure 6.1 presents a sketch of the Crimean Peninsula.

The situation for Ukraine changed when the Cold War ended (1989), and the Soviet Union collapsed. In 1990, Boris Yeltsin and Leonid Kravchuk signed a treaty in Kyiv in which both leaders recognized the sovereignty of both states. When a

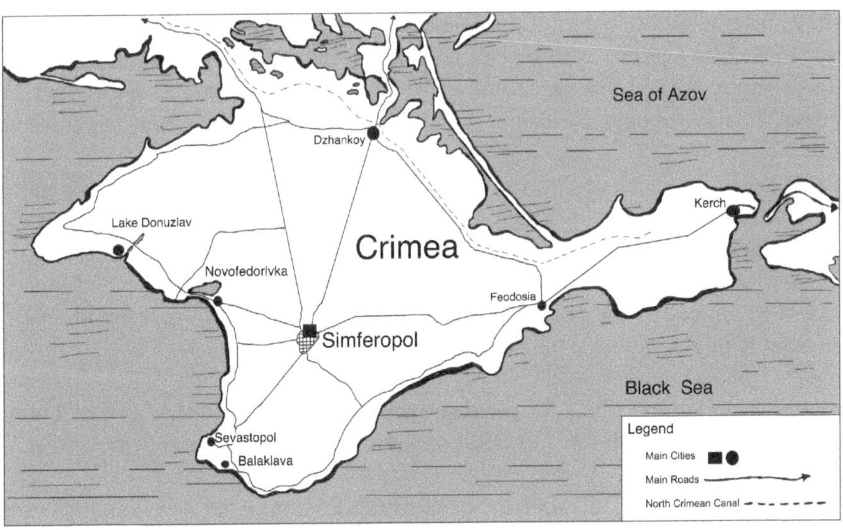

Figure 6.1 The Crimean Peninsula.

referendum was held on December 1, 1991, 90 percent of the Ukrainian population voted for independence. While in 25 of the 27 Oblasts, between 83 and 97 percent of the population voted for independence (turnout 70 percent), only between 54 and 57 percent did so in the Crimean Autonomous Soviet Socialist Republic (ASSR) and Sevastopol City (turnout 37 percent).[47] These numbers illustrated the pro-Russian attitude of the population in these two areas. A week later, Yeltsin recognized Ukraine's independence unconditionally. The Crimean Peninsula became the Autonomous Republic of Crimea (ARC) in the newly established and independent Ukrainian state. Russia, however, kept a vital interest in the peninsula, as the city of Sevastopol was home to its Black Sea Fleet. Without this naval base, Russia would lose its direct access to the Mediterranean Sea. To prevent such loss, both countries signed the Partition Treaty on the Status and Conditions of the Black Sea Fleet in 1997, which allowed Russia to continue using Sevastopol to harbor its fleet until 2017.[48] In 2010, the agreement between Ukraine and Russia on the Black Sea Fleet in Ukraine, also known as the Kharkiv Pact, replaced the Partition Treaty and gave Russia continued access to the port until 2042.[49] The treaty allowed Russia to station up to 25,000 troops at the naval base. Any Russian troop movements outside the base had to be reported to the Ukrainian authorities in advance. Russian troops were also not to interfere in Ukrainian domestic affairs.

Planning and preparation

On January 30, 2014, Konstantin Malofeev, a Russian businessman with close ties to Putin, joined a Russian religious delegation that visited the Ukrainian capital, Simferopol. Igor Girkin, a former officer of the FSB who became Minister of Defense in the Donetsk People's Republic (DPR) during the Donbas War, was found guilty of his role in the shootdown of Malaysia Airlines Flight 17, which led to the death of 298 people, was also part of the delegation. During the trip, both men conducted reconnaissance and met with several leading figures of the pro-Russian separatists. Upon their return to Moscow, Malofeev wrote a memo to Putin recommending that Russia should take large parts of eastern Ukraine, including the Crimean Peninsula:

> In these circumstances, it seems appropriate to play along the centrifugal aspirations of the various regions of the country, with a view to initiate the accession of its eastern regions to Russia, in one form or another. Crimea and Kharkiv region should become the dominant regions for making such efforts, as there already exist reasonable large groups there that support the idea of maximum integration with Russia.[50]

Vladislav Surkov, Russia's former deputy prime minister and then one of Putin's closest advisers, conducted an additional reconnaissance mission to Crimea in early February.

On February 20, 2014, when the demonstrations at *Maidan Nezalezhnosti* escalated and led to the death of a substantial number of demonstrators and police officers, Putin's concern began to grow. When Yanukovych decided to flee the country two days later, Putin told some of his trusted aides "that the situation in Ukraine has unfolded in such a way" that it forced him to decide "to begin the work of returning Crimea to Russia."[51] By now, the planning of the annexation was completed, and he ordered his special services to exfiltrate Yanukovych to Russia.[52] According to Putin, the Russian forces had to protect the Russian-speaking majority of Crimea from violent Ukrainian nationalists and to defend the Russian troops based at Sevastopol against aggression from the Ukrainian military.[53] Due to the unfavorable balance of forces, according to Putin, the operation required specially selected and highly skilled personnel able to operate covertly. Therefore, Putin ordered the Russian ministry "to deploy special forces of the GRU (military intelligence) as well as marines and commandos there under the guise of reinforcing security for our military facilities in Crimea."[54] Shortly after that, GRU operatives began to shape the environment by coordinating with pro-Russian separatists about their role in the operation.[55]

The actual annexation

February 25

The first signs of the start of the operation became visible on February 25, when thousands of pro-Russian citizens gathered in Sevastopol to demonstrate against the Maidan Revolution, demanding greater Crimean autonomy from Ukraine and seeking closer ties with Russia. Following the resignation of Volodymyr Yatsuba as mayor of Sevastopol, the city council decided to install Alexey Chaly, a Russian citizen, as his successor.[56] At the same time, Russian military units belonging to the 810th Naval Infantry left their base in Kazachya Bay, in the vicinity of Sevastopol, in Armored Personnel Carriers (APCs) and took positions at key intersections leading to the city. These actions were the first Russian violation of the rules on troop movements outside of its bases that always required a police escort, as described in the Kharkiv Pact. When approached by Ukrainian forces, the Russians informed them that they were protecting Russian military installations against aggressive actions from fascists.[57]

Additionally, unmarked self-defense militias began to block the access gates to the Ukrainian Navy Headquarters in Sevastopol. While these forces stated that they were defending the headquarters from violent Ukrainian far-right groups (*Pravy Sektor*), they were essentially isolating the installation and thereby disrupting Ukrainian command-and-control.[58] Finally, similar forces occupied blocking positions surrounding the Saky airbase at Novofederivka, approximately 70 kilometers north of Sevastopol and home to the 10th Naval Aviation Brigade.[59] This specific action was meant to defend the base against ethnic Tatars (a minority group of Sunni Muslims) who, according to the self-defense forces, intended to seize the base and its weaponry to harass the Russian-speaking population.[60] This claim was later repeated by several senior Russian military officers who came to the base to negotiate about transferring the Ukrainian weapons to a secure location to reduce the risk of them being seized by the Tatars in case they raided the base.[61]

The same day, the *Nikolai Filchenkov*, a Tapir Alligator class amphibious landing ship, arrived at the Sevastopol Naval Base.[62] On board the vessel were approximately 200 Russian SOF (*Sily spetsial'nykh operatsii*, or SSO) belonging to the recently established Special Operations Forces Command (*Komandovanie sil spetsial'nalnykh operatsii*, or KSSO).[63] According to other Ukrainian sources, multiple ships belonging to the Black Sea Fleet but based in Novorossiysk arrived at the Sevastopol Naval Base that same day. They reported to the State Border Guard Service (SBGS) of Ukraine that they were not carrying any cargo.[64] Such movements of ships were permitted under the Kharkiv Pact. Ukrainian military reporting, however, suggests that approximately 1,000 Russian VDV troops were on board of these ships.[65]

February 26

The next day, on February 26, Russian television began to strongly influence the Crimean population by claiming that the interim government in Kyiv was illegitimate, the Russian-speaking population in Crimea was under threat, and Russia had to act to protect them.[66]

At the same time, the Kremlin announced a snap inspection of the combat readiness of 150,000 troops assigned to the Western and Central Military District Command Posts.[67] The Kremlin further announced that Putin and Defense Minister Sergey Shoigu would personally attend the exercise.[68] Although inspections of this scale are not uncommon in the Russian military, the purpose of this event was to conceal the movement of a large number of Russian *Spetsnaz* and elite airborne troops (*Vozdushno-Desantnye Voyska*, or VDV) by approximately 40 Ilyushin Il-76 military transport planes to a staging area in the vicinity of the Kerch Strait that separates Russia from Crimea and connects the Black Sea with the Sea of Azov.[69]

The massive exercise caught the attention of the Ukrainian National Security and Defense Council (NSDC), which assessed that this was an act of deception (*maskirovka*) and that an invasion was looming.[70] As a result, the attention of the NSDC and, more importantly, intelligence resources, were directed to monitoring the exercise along Ukraine's eastern border rather than the events that were about to unfold in Crimea.[71]

February 27

On February 27, events started to speed up and grow in significance. A total of approximately 50 Russian SOF from KSSO, disguised as a local self-defense militia, seized the buildings of the *Verkhovna Rada* of Crimea and the ARC government in Simferopol without firing a shot. This was possible because the local police facilitated their entry, as recorded by video surveillance. Shortly thereafter, the Russian SOF raised the Russian flag over the building.[72] Because of their disguise, which included unmarked green army uniforms (hence the popular term "little green men"), the operators could not be attributed to Russia. Moreover, the analysis of several photographs taken during the seizure of the building indicates that Russian VDV troops pretended to be legitimate security forces by wearing the official uniforms of Ukrainian police units based in Crimea.[73] These deceptive actions strengthened Russia's denial of involvement, as the photos reflected the participation of unmarked self-defense forces assisted by the Ukrainian police and allowed its leadership to deny their involvement. For example, when Euronews asked Vladimir Chizhov, the Russian Ambassador to the EU, about Russia's role in the events, he answered: "There are no troops whatsoever, no Russian troops at least."[74] Putin was also asked about the unmarked troops, whose uniforms, according to a reporter, looked Russian. In his response, Putin emphasized, "Look at the post-Soviet states. There are many uniforms there that are similar. You can go to a store and buy any kind of uniform."[75]

Following the seizure of the buildings in Sevastopol, the ARC government was dissolved and reconstituted under the supervision of the Russian separatists, who voted to organize a referendum on May 25 to gain more autonomy from Kyiv.[76] Next, to vote on the date for the referendum, nearly 90 percent of the representatives (55 of 64) elected the Russian politician Sergey Aksyonov as the new prime minister.[77] At the same time, another large Russian naval vessel arrived in Sevastopol harbor, carrying an additional number of 300 Russian troops.[78] Although the Russians followed the appropriate procedures for this troop movement, they failed to inform the Ukrainians in advance, thereby again violating the agreements made in the Kharkiv Pact. By the end of the day, unmarked Russian-speaking men also surrounded a military airfield in Belbek, just north of Sevastopol and home to the 204th Tactical Aviation Brigade.[79]

February 28

More apparent signs of Russian military involvement became visible on February 28, when three Russian Mi-8 medium transport helicopters accompanied by eight Mi-35M (Hind E) attack helicopters entered Ukrainian airspace unauthorized.[80] Before the Ukrainian Air Force could respond effectively, the helicopters were already flying over the peninsula to support Russian ground forces. The Russians also seized the International Airport in Simferopol, the second-largest city of Crimea. All civilian flights were canceled, which further isolated the peninsula and provided the Russians the ability to bring in additional troops by air. Moreover, Russian forces in disguise seized the peninsula's main television tower as well as the state television and radio channel GTRK *Krym* in Simferopol, thereby effectively taking control of all Ukrainian channels.[81]

March 1

Up to this point, the Russians were still numerically inferior to the Ukrainians. This inferiority changed when several Russian amphibious landing ships brought a substantial number of forces and equipment to the peninsula on March 1. By doing so, Russia violated the agreement made in the Kharkiv Pact for a third time, as the number of ships now exceeded the authorized number.[82] Equipped with light tactical vehicles and APCs, these forces quickly seized a sizable portion of territory, including military bases and other key administrative buildings. At the same time, Aksyonov moved the date of the referendum up from May 25 to March 30.[83]

The Russians also attempted to gain support from the local Ukrainians by conducting false flag operations, using propaganda and psychological pressure. For example, Russian SOF disguised as Ukrainian nationalists pretended to conduct a raid on the building home to the Council of Ministers in Simferopol on March 1. Pro-Russian separatists and unmarked VDV troops, however, were already controlling the building at that time. Various sources provide evidence to support the fact that this was a false flag operation. First, there were almost no hostilities observed, no casualties reported, no arrests made, and the attackers stayed on the scene for only a limited amount of time.[84] Furthermore, some observed that the attackers were equipped with GM-94 grenade launchers, a weapon specifically built for Russia's special services and SOF.[85] Both Russian and Western media covered the raid, which led to broad news reporting on aggressive Ukrainian nationalists. This media coverage strengthened Russia's narrative that the Russian-speaking population was at risk.

March 2

A potentially significant event occurred on March 2, when Rear Admiral Denis Berezovsky, Commander of the Ukrainian Navy, defected to the Russian separatists. In a speech given on March 3, he called upon his subordinate officers to follow his example. He was unsuccessful, however, as most of them remained loyal to Ukraine at that time.[86] The same day, Vitaly Churkin, Permanent Representative of the Russian Federation to the United Nations (UN), presented a letter during a Security Council (SC) meeting in which he claimed that Yanukovych (who had fled to Russia) requested Russia's assistance to restore peace and order in his country. The letter was signed on March 1 and included the following:

> As the legitimately elected President of Ukraine, I wish to inform you that events in my country and capital have placed Ukraine on the brink of civil war. Chaos and anarchy reign throughout the country [...] I therefore call on President Vladimir Vladimirovich Putin of Russia to use the armed forces of the Russian Federation to establish legitimacy, peace, law and order, and stability in defence of the people of Ukraine.[87]

The same day, the Russian Federation Council approved the deployment of military forces to Ukraine. In a press release, Putin argued that this was a legitimate action because it was at the request of the legitimate President of Ukraine.[88] In other words, Putin claimed that he was helping an ally with the defense of his country rather than intervening out of his own interests.

March 6

On March 6, the separatists changed the date of the referendum for a second time, moving the vote up from March 30 to 16.[89] Importantly, the separatists announced that they had voted to become part of Russia, thereby changing the purpose of the referendum as well. Additionally, they requested Russia to start the procedure to officially transfer Crimea to Russia.[90] In the days prior to the vote, Putin seems to have influenced these changes by publicly stating that although becoming part of Russia was not a viable option at that time, the Crimean population had similar rights as the Albanians had in Kosovo.[91]

March 7–15

In the following days, Russia began to add conventional forces to the operation. First, a substantial buildup of forces was observed in the area near the Kerch Strait, where VDV troops had prepared before their deployment to the peninsula. This troop buildup consisted of motorized units, artillery batteries, and air defense units, as well as anti-ship missile batteries.[92] Additionally, Russian forces were deployed to the Chonhar Strait, which separates mainland Ukraine from the Crimean Peninsula. By doing so, the Russians effectively isolated the peninsula from the air, land, and sea.

With the peninsula isolated and the buildings of the *Rada* and ARC government, as well as key military installations either seized or blocked, Ukrainian command-and-control was paralyzed. To further disintegrate the Ukrainian ability to organize an effective response, the Russians cut off landline communications through classical sabotage and employed Electronic Warfare (EW) to jam cellphone and radio signals. For example, the phone and Internet cables belonging to the Ukrainian state-owned telecommunications service provider Ukrtelecom were targeted. The Russians also conducted cyber-attacks against Ukrainian government websites. As a result, cellular or Internet communications between the peninsula and the Ukrainian mainland became disrupted.[93] Finally, Russian troops took nine Ukrainian television channels from the air.[94] This action left the Crimean population with Russian-speaking and Russian-state-owned channels that allowed Russia to influence the Ukrainians by broadcasting propaganda and spreading disinformation.

The Russians were aware that the events on the peninsula could lead to civil unrest. Therefore, GRU operatives had been responsible for preparing the environment immediately after Putin decided to launch the operations. A vital part of these preparations consisted of recruiting local militias.[95] The Cossacks, a group of people predominantly consisting of East Slavs and familiar with the area, appeared to have been working closely with the Russians. Members of the *Berkut* (special police) also came to Crimea after their unit was disbanded because of their role in the violent escalations at *Maidan Nezalezhnosti* and joined the separatists.

The presence of the Russian forces, particularly the disguised ones, strongly boosted the Russian separatists' morale and confidence in their ability to support the Russian objective. For example, the separatists feared a potentially powerful Ukrainian response to the protests and election of Chaly as the new mayor of Sevastopol. The supportive attitude of the disguised Russian forces gave them the necessary psychological support to become more active as the events unfolded.[96]

March 16

On March 16, the separatists organized the referendum. The referendum included two questions, one of which the voters had to answer with yes:

1 Are you for the reunification of Crimea with Russia as a subject of the Russian Federation?
2 Are you for the restoration of the actions of the 1991 Constitution of the Republic of Crimea and for the status of Crimea as part of Ukraine?[97]

These two questions indicate that the voters could not choose for a continuation of the existing situation, as described in the 1998 Crimean Constitution.[98] It should be stressed that simply voting with a "no" was therefore not an option.

March 17

The results of the referendum were published the next day. According to Mikhail Malyshev, the head of the Crimean referendum commission, a total of 1,274,096 people (turnout 83.1 percent) voted; 1,233,002 (96.8 percent) favored unification with Russia.[99] The Sevastopol City Council also published the results of the referendum in the city itself. According to them, a total of 274,101 (turnout 89.5 percent) people voted, of which 262,041 (95.6 percent) voted to become part of Russia.[100] Next to these local news reports, Russian state media also published the results of the referendum, thereby broadly articulating the apparent desire of the Crimean population to become part of Russia.[101] Western news agencies did so, too, but found the referendum illegitimate.[102] By the end of the day, the separatists issued a declaration of independence that articulated that the outcome of the referendum was "based on the direct expression of the will of the peoples of the Crimea."[103]

March 18

On March 18, Putin signed an executive order that stated the following:

> Given the declaration of will by the Crimean people in a nationwide referendum held on March 16, 2014, the Russian Federation is to recognize the Republic of Crimea as a sovereign and independent state, whose city of Sevastopol has a special status.[104]

With that, Russia had successfully completed the operation without firing a shot and taken Crimea by *fait accompli*.

In the days following the official annexation, Russian conventional military forces occupied most of the military bases and installations on the peninsula. With a few exceptions, they received no resistance from Ukrainian forces. To make Ukrainian forces shift sides, they were offered Russian citizenship accompanied by a position in the Russian army while honoring rank and also receiving better pay. Most Ukrainians did so because of the shared language, culture, and ethnicity. Practical reasons also drove choices, as many Ukrainians preferred to remain stationed on the peninsula, close to their families. Out of the 18,000 Ukrainian soldiers stationed in Crimea, only one-third decided to depart to mainland Ukraine, while even a portion of those later returned to the peninsula.[105] By March 26, the operation was completed, and Russia started to move its troops back to the homeland.

Throughout the entire operation, Russia strongly denied it was behind the self-defense militias. It was only by April 17 that Putin admitted Russian forces played a limited role in the operation by supporting the self-defense forces. Russia's entire role was finally revealed in March 2015, a year after the annexation. In a documentary titled *Crimea: The Way Home*, Putin provided a considerable number of details of the operation.[106] With Putin's confession, the annexation of Crimea was officially attributed to Russia.

Analysis

Policy

The objective of making Crimea part of Russia was something that had not left the minds of many Russian elites. As the situation in Ukraine was becoming a threat to Russia's national interests, Putin decided it was time for swift action. One of the triggers was the Ukrainian *Rada's* decision to ban Russian as Ukraine's second language. The Maidan Revolution and the subsequent flight of Yanukovych also demonstrated that Russia was quickly losing its control over Ukraine. These events, coupled with the severe risk posed by a potential shift to the EU and perhaps even NATO, made Putin decide "to return Crimea to Russia."[107]

Military strategy

Given the dynamic events that took place in early February, coupled with the short response time, Russia was probably unable to devote much time to developing the strategy to achieve its political objective. Reflecting on the Soviets' historical successes in Budapest (1956), Prague (1968), and Kabul (1979), Putin decided that Russia would annex Crimea through a classical *fait accompli* strategy.

Russia had to achieve five complementary objectives to successfully present the Ukrainians (and the West) with a *fait accompli*.

1 GRU operatives had to support the pro-Russian separatists with the planning, preparation, and execution of demonstrations against the current Crimean (i.e., Ukrainian) authorities.
2 SOF had to seize the building of the Crimean *Rada* to allow for the election of Aksyonov as new prime minister, vote for the referendum and ask Russia to start the official procedure to officially transfer Crimea to Russia.
3 VDV troops would block key military bases and administrative centers as well as occupy the Simferopol International Airport. At the same time, conventional

Russian forces would establish blocking positions at the Kerch Strait, Chonhar Strait, and the Sevastopol Naval Base, thereby isolating the peninsula from mainland Ukraine.
4 Russian forces would sabotage land communication lines, employ EW, and conduct cyber operations to disrupt communications between Crimea and mainland Ukraine. These actions, coupled with propaganda, would influence the Crimean population to support the Russian separatists in their claim to become part of Russia.
5 The separatists had to organize the referendum, the outcome of which would make the annexation official.

To assess the inherent logic of the Russian strategy, the following sections will describe its suitability, feasibility, and acceptability.

Suitability: could the Russians execute the operation with the available resources?

The composition of the forces played a decisive role in the successful outcome of the operation. As directed by Putin, Russia employed GRU operatives and VDV troops during the initial and critical phase of the operation. Moreover, Russian SOF from the newly established KSSO was assigned to seize high-value targets such as the buildings of the *Rada* and ARC government. By doing so, Russia employed its most specialized and highly trained forces that could operate below the detection or attribution thresholds and achieve all five ends without any significant challenges.

Four strategic factors turned out to be vital for the successful outcome of the operation.

1 The Crimean Peninsula has unique geographical characteristics that make it relatively easy to isolate. By controlling the Kerch Strait to the east, Chonhar Strait to the north, establishing a blockade at the naval base in Sevastopol, and seizing the international airport as well as several military airfields, Russia was able to quickly isolate the peninsula from the air, land, and sea. By sabotaging land communication lines, jamming cellphone signals, and conducting cyberattacks on Ukrainian government websites; the same applied to cyberspace and the Electromagnetic Spectrum (EMS). In other words, Russia could isolate Crimea in all domains of military operations. Crimea also borders on Russia, allowing it to use its own territory to make the required preparations for the operation, including pre-positioning military forces. The same applied to the fact that both Russia and Ukraine stationed a substantial number of troops on the peninsula, something that both countries agreed upon in the Kharkiv Pact and was accepted by the local population. From that perspective, the naval base in Sevastopol stands out, as the agreement allowed the Russians to use the base to bring in a substantial number of forces undetected.
2 Russia and Crimea have a shared language, culture, and ethnicity. This proximity enabled GRU operatives to cooperate effectively with the leadership of the

pro-Russian separatists. For example, the GRU could infiltrate the area undetected and blend in among the protesters, or self-defense force units. Moreover, Russian SOF and VDV troops could disguise themselves as self-defense forces to seize key objectives or pretend to be local police forces to conduct riot control against the opposition.

3 Next to the shared language, culture, and ethnicity, the Russians exploited the strong affinity (*bratstvo*) many Ukrainian military personnel still had for their Russian counterparts. These feelings originated from the fact that senior military personnel from both countries had served together in the Soviet Army. Also, both navies had jointly patrolled the Black Sea. Because of these close ties, personnel from both sides frequently met with each other, played sports together, and attended each others' festivities, not to mention that they lived alongside each other on the peninsula. These close ties made many Ukrainian service members believe the Russian claims of protecting the population and military installations against impending attacks from fascists or violent ethnic Tatars rather than supporting an ongoing annexation. The close ties between the Russians and Ukrainian service members also facilitated the talks surrounding the takeover of military installations that were completed without any violence.
4 The transit agreements between Russia and Ukraine were a final strategic factor. The Kharkiv Pact allowed Russia to keep up to 25,000 troops on the peninsula. To rotate personnel in and out as well as to supply its naval base in Sevastopol, Russia was allowed to use airplanes and ships. At the beginning of the operation, this permission enabled Russia to bring in the first wave of troops, including SOF and VDV troops, undetected. At a later stage, Russia brought in additional airplanes, ships, and troops, thereby violating the agreement. Although detected by the Ukrainians, they found they could not stop these Russian actions at that moment.

Feasibility: did achieving the objectives create the strategic effects that contributed to the desired end?

Russia achieved all five objectives described above. By doing so, Russia created the strategic effects that contributed to the desired end of returning Crimea to Russia. For example, the GRU was successful in providing support to the pro-Russian separatists. Of significance was the employment of clandestine Russian VDV troops that boosted the separatists' confidence in their potential success. Second, the Russian SOF seized the buildings of the *Rada* and the ARC government without firing a shot. By controlling the *Rada*, Russia enabled the separatists to elect Aksyonov as their new prime minister, to vote favorably for the referendum, and to request Russia to start the procedure to annex Crimea. Third, by effectively blocking key military bases and administrative centers, seizing the international airport at Simferopol, and establishing effective blocking positions at the Kerch Strait, Chonhar Strait, and naval base at Sevastopol, Russia effectively isolated Crimea from mainland Ukraine. Fourth, through effective sabotage of land communication lines, targeted EW, and computer network operations, coupled with propaganda, Russia cut off all communications with the outside world and influenced the local population to support their intended objective. Finally, integrating all four effects allowed the separatists, with Russian support, to organize the referendum. The achievement of this final objective created the strategic effect of a majority vote for the option to become part of Russia. When the separatists, based on the outcome of the referendum, issued a declaration of independence, Crimea became part of Russia, and Putin had achieved his desired end.

Acceptability: did the strategic effects sought to justify the objectives pursued and consider tangible and intangible costs?

To determine whether the objectives meet the condition of acceptability based on consequences for Russia, it is crucial to look at two distinct forms of cost: costs resulting from Ukraine's retaliation and costs resulting from an international response.

During the execution of the operation, the Ukrainian government was dysfunctional. The recent resignation and subsequent flight of Yanukovych in response to the escalating protests at *Maidan Nezalezhnosti* meant that the government was faltering. Decision-making was hampered by the chaos caused by the violent protests. During NSDC meetings, the uncertainty about what was unfolding in Crimea, coupled with the relatively deficient interim Ukrainian government, led to reluctance and unwillingness to deliver an effective response. The risk of escalation also played a part herein.[108] Furthermore, the Head of Ukraine's security service (*Sluzhba Bezpeky Ukrayiny*, or SBU) reported that the morale of the military and security forces was very low and that they would not be receptive to orders from the interim government.[109] The greatest concern in this regard relates to the naval forces based in Sevastopol: the authorities believed they would defect to Russia.

In sum, the Ukrainian government was paralyzed and unwilling to respond militarily. Even if they had decided to do so, it is doubtful if the Ukrainian military would have executed the orders, given its low morale and distrust in the interim government. Russia had thus nothing to fear from Ukrainian retaliation.

Just as the speed of events had surprised Putin, so too should it have surprised the West. It is, therefore, plausible that neither the EU nor the US had started any contingency planning when the situation unfolded. A military intervention was not on the table, as Ukraine is not a member of NATO, so Article 5 was not applicable. It is even doubtful if the West could have deterred Russia from conducting the operation. Deterrence by denial would certainly not have been a viable option. Punishment with economic sanctions might have, although the West's dependence on Russian oil and gas probably would have prevented reaching a consensus among the EU member states. With that, the door for Russia to launch the operation and secure the gains it hoped to make was wide open, while the risk of tangible or intangible costs was low.

Risk

What follows is an assessment of the risks associated with the Russian strategy. This assessment includes looking at the consequences of success and failure, examining the ends, ways, and means relative to the strategic environment, and determining which factors impacted the generation of strategic effects. Since this chapter focuses on the annexation of Crimea, which took place in February and March of 2014, the risk assessment is focused on the immediate consequences of that operation rather than the conflict in Donbas that started in April 2014.

What assumptions did the Russians make in their strategy, and what was the effect of any wrong assumption?

Russia made several assumptions in its strategy that proved correct. For example, the Russian belief that the shared language, culture, and ethnicity were exploitable allowed them to successfully recruit and support pro-Russian separatists. They could also exploit the strong affinity (*bratstvo*) the Russians assumed many Ukrainian service members still had for them. As a result, the Ukrainians believed the Russian claims of protecting the population and military installations against attacks from fascists or violent ethnic Tatars rather than supporting an ongoing annexation. The close ties also facilitated the talks surrounding the takeover of military installations that were mostly completed without any violence. Finally, the Russians strengthened the unsympathetic attitude they believed the Crimean population already had toward Kyiv. By doing so, the Russians could manipulate the outcome of the referendum to their benefit.

The Russian strategy also included several assumptions that proved to be incorrect. The effects of these incorrect assumptions, however, turned out to be of little influence on the outcome of the operation. For example, the Russians could not make the Ukrainian military defect on a large scale. What stands out is the defection of Rear Admiral Denis Berezovsky, Commander of the Ukrainian Navy, on March 6. However, his defection proved insignificant, as only a limited number of Ukrainian service members followed his example. Although a substantial number of Ukrainian troops defected after the completion of the annexation, these defections did not affect the outcome of the operation anymore.

Did the Russians incorporate flexibility or adaptability into their strategy? If required, how could they have modified their strategy, and at what cost?

Two components of the Russian strategy seem to have offered flexibility and were changed throughout the operation. The first flexible component was the date of the referendum, which the separatists adjusted twice. More importantly, the separatists changed the purpose of the referendum itself. Putin seems to have ordered these adjustments when he observed that the operation was proceeding according to the plan and no major setbacks were experienced. For example, Russia could have encountered Ukrainian resistance, domestic public discontent, or pressure from international responses. When none of these occurred, the date of the referendum, initially scheduled for May 25, was moved to March 30 and ultimately even to March 16.

The second flexible component was the flexibility regarding the cunning approach to the local Crimean leadership. At the beginning of the operation, Russia asked Anatoly Mogilev, the existing Crimean prime minister, to be the leader of the pro-Russian separatists. When he refused, Russia looked into the possibility of recruiting Leonid Grach. When the Russians discovered that public support for him was lacking, they dropped him. The third option was Aksyonov, who stood out during the clashes with ethnic Tatars in Simferopol. After the successful seizure of the building of the *Rada*, he was elected prime minister.

How did other actors react to the Russian invasion?

The annexation of Crimea held significant implications for countries that were formerly part of the Soviet Union, including the Baltic states. According to Putin's press statements and Lavrov's responses to questions in the international media during the operation, Russia believed it had the right to intervene in sovereign states to protect Russians who, unfortunately, lived outside Russia's current borders. Next to Crimea, similar coercive arguments were previously used to station military forces in Transnistria (Moldova) or occupy parts of Georgia. That the countries in Russia's "near abroad" had to take the Russian threat seriously became clear in September 2014, when Putin told Ukrainian President Poroshenko that "in two days I could have Russian troops not only in Kyiv, but also in Riga, Vilnius, Tallinn, Warsaw, and Bucharest."[110]

The Western response was considered weak and received a lot of criticism. For example, the American president, Barack Obama, was blamed for his failure to recognize Putin's increasing authoritarianism and aggressive posture toward the countries in Russia's "near abroad."[111] Additionally, Obama's unwillingness to deliver a powerful response when Syrian president Bashir al-Assad crossed his "redline" by using chemical weapons, arguably strengthened Putin's belief that he could get away with the operation.[112] Malcolm Rifkind, the former British foreign secretary, described the Western response as follows:

> On the basis of the measures announced so far by both the US and the EU, on visa controls and asset freezes internationally, I say with great sadness that is a pathetic and feeble response that does not match the seriousness which those implementing these responses have themselves acknowledged we face at the present time.[113]

One course of action of the West was to discuss the annexation in the UN Security Council (UNSC). The draft resolution stated that the members of the UNSC reaffirmed Ukraine's "sovereignty, independence, unity, and territorial integrity." The members also condemned the referendum and its outcome as illegitimate. As a permanent member of the UNSC, Russia would, of course, use its veto. Therefore, the meeting was to demonstrate Russia's isolated position. Ultimately, the General Assembly passed a non-binding resolution. Despite its non-binding character, the resolution was passed with one hundred votes in favor, eleven against, and 58 abstentions.[114]

The fact that Ukraine did not receive any support from the West could lead to skepticism over the Nuclear Non-Proliferation Treaty (NPT). Ukraine signed the NPT in December 1994. By doing so, giving up on "1,240 nuclear warheads arming 176 intercontinental ballistic missiles (ICBMs), including their extensive launch control infrastructure, 700 nuclear cruise missiles arming 44 strategic bombers, and nearly 3,000 tactical nuclear weapons, including artillery shells, gravity bombs, and mines."[115] The fact that none of the major nuclear powers (i.e., China, France, the US, and the UK) supported Ukraine could lead to growing skepticism about the treaty's credibility.

How did chance or friction play in this strategy?

The operation was conducted without any unforeseen obstacles. In fact, many military planners, security experts, and members of the intelligence community credited Russia with the execution, as its forces successfully exploited several critical Ukrainian vulnerabilities.[116] Thus, there are no indications that chance or friction played a role in this strategy.

Tactics

Did the Russians achieve relative superiority throughout the operation?

The Russians achieved relative superiority by applying the principles of security, speed, surprise, and legitimacy throughout the operation. During the planning and preparation phases of the operation, Russia's special services were able to successfully recruit and support the pro-Russian separatists undetected. The Russians could also insert a substantial number of SOF and VDV troops, required for the annexation, into Sevastopol Naval Base undetected by exploiting the agreements made in the Kharkiv Pact. These clandestine preparations allowed the Russians to gain an advantage before they reached the point of vulnerability. By doing so, they reduced the area of vulnerability by half. Figure 6.2 illustrates this by the point of vulnerability starting above the Z-axis (i.e., relative superiority line) and allowing the Russians to enter the engagement with relative superiority. Having achieved relative superiority at this early stage in the operation meant that the Russian chance of mission success outweighed the chance of mission failure.

During the initial phases of the engagement, the Russians further strengthened their relative superiority by applying the principles of speed and surprise. While disguised as local self-defense forces, Russian SOF quickly seized several high-value targets, including the buildings of the *Rada* and ARC government, which allowed them to set the critical conditions for the annexation, including the election of Aksyonov as new prime minister, vote for the referendum and ask Russia to

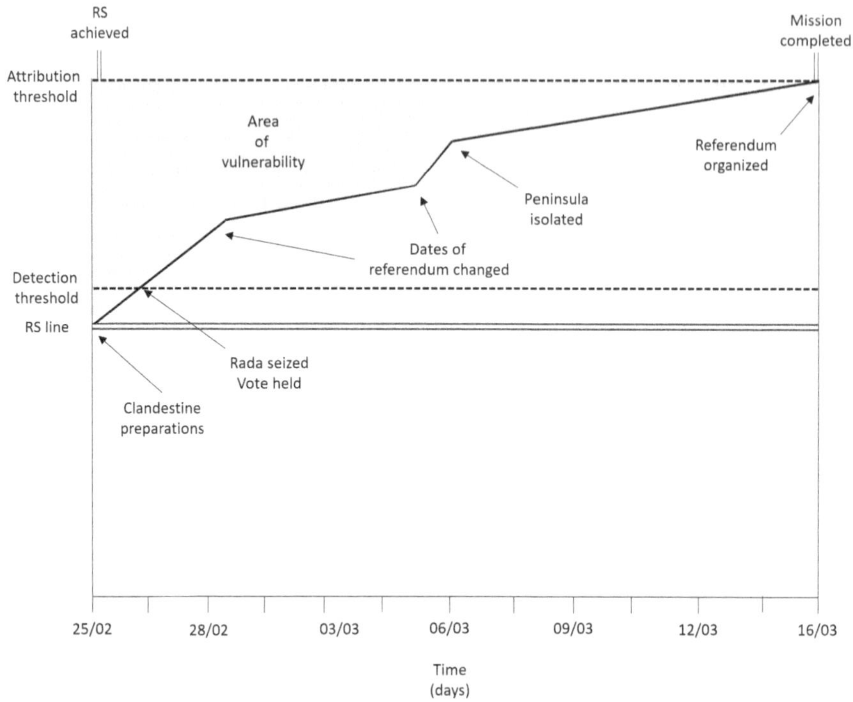

Figure 6.2 Relative superiority graph for the Russian annexation of Crimea.

start the procedure to officially transfer Crimea to Russia. At the same time, VDV troops, wearing disguises, rapidly isolated several key Ukrainian military bases, and, thereby, disrupting Ukrainian command-and-control. These successful acts of deception allowed the Russians to operate below the attribution threshold and deny their involvement in the events. At this point in time, the chance of mission success was even further outweighing the chance of mission failure. Figure 6.2 illustrates this by the steep curve that reaches far above the Z-axis, thereby substantially decreasing the vulnerability area.

If achieved, how did the Russians sustain their relative superiority throughout the operation? If not, why did they fail?

The Russians could sustain their relative superiority by applying the principles of speed, surprise, security, and legitimacy. The successful seizure of the building of the *Rada* by the Russian SOF enabled the separatists to elect Aksyonov as their new prime minister, to vote favorably for the organization of the referendum, and request Russia to start the procedure to annex Crimea. The mission was not completed with these actions. For that to be achieved, the referendum itself had to be held. With the referendum initially scheduled for May 25, the Russians had to retain control over the Ukrainian military and local population as well as prevent any outside interventions from either Ukraine or the international community while remaining unattributed. Thus, assuming the informal annexation on February 27 (i.e., the election of a new prime minister by the reconstituted Crimean government), the originally anticipated time frame for the referendum was shortened from approximately two months to just two weeks. Therefore, the Russians had to sustain the condition of relative superiority until the outcome of the referendum was known.

As illustrated by Figure 6.2., the Russians could sustain their relative superiority until mission completion by further decreasing the area of vulnerability. They achieved this by continuing the employment of deception at the tactical and strategic level and applying the principle of security while simultaneously using speed to move the date of the referendum up to March 16. The Russians also successfully applied the principle of legitimacy, which enabled them to exploit Ukrainian vulnerabilities, including their shared language, culture, and ethnicity, as well as the strong affinity (*bratstvo*) that many Ukrainian troops still had for their Russian counterparts. The combination of these principles proved to be vital for successfully sustaining relative superiority until the separatists could organize the referendum and announce its outcome, thereby making the annexation official.

Conclusion

The annexation of Crimea is a clear example of how superior tactical action can lead to strategic effect. The initial Russian force was small and consisted of GRU operatives, SOF, and VDV troops, augmented with infantry units. This small force paralyzed a numerically superior force of 18,000 equipped with heavy firepower and remained one step ahead of them from start to finish.

The Russians were successful because they selected their best personnel, capable of exploiting specific Ukrainian vulnerabilities. The Russians could seize the peninsula, contain, and ultimately disarm a numerically superior force, control, and influence a population of two million people while operating below the attribution threshold. By doing so, Russia's political leaders and military commanders demonstrated a high level of strategic sense, as they well understood how to convert tactical performance into strategic effect. The case, therefore, illustrates one in which the Russians fulfilled the strategic promise of special operations through the careful alignment of policy, strategy, and tactics.

Notes

1 Serhii Plokhy, *The Gates of Europe: A History of Ukraine* (New York: Basic Books, 2015).
2 The Avalon Project, "Foreign Relations of the United States: 1918 The Conclusion of the Peace of Brest Litovsk The Peace Treaty of Brest-Litovsk 1918," Yale Law School, 2008.
3 Samuel Kucherov, "The Future of the Soviet Collective Farm," *American Slavic and East European Review* 19, no. 2 (1960): 180–201.
4 Plokhy, *The Gates of Europe*.
5 Verkhovna Rada Ukraine, No, 1427-XII, Act of Declaration of Independence of Ukraine," August 24, 1991.

6 Transparency International, "Corruptions Perceptions Index."
7 Peter Finn, "Widespread Vote Fraud Is Alleged in Ukraine," *The Washington Post*, November 23, 2004.
8 Lili Bivings, "Ukraine's Orange Revolution," *The Kyiv Independent*, August 2, 2023.
9 Wilson Center, "Orange Revolution: Ukraine Votes for Change," *Kennan Institute*, March 8, 2005.
10 Elisabeth Rosenthal, "Liberal Leader from Ukraine Was Poisoned," *The New York Times*, December 12, 2004.
11 The Associated Press, "Ukrainian President Fires His Government," *The New York Times*, September 8, 2005.
12 Peter Finn, "Ukrainian President Turns to A Rival," *The Washington Post*, September 23, 2005.
13 Christian Caryl, "Yanukovych Won. Get Over It," *Foreign Policy*, February 8, 2010.
14 Ellen Barry, "Former Ukraine Premier Is Jailed for 7 Years," *The New York Times*, October 11, 2011.
15 Dimitri Medvedev, "Dmitry Medvedev Congratulated Viktor Yanukovych on his Victory," *The Kremlin*, February 15, 2010.
16 Martin Grančay, Ērika Šumilo, and Jolita Vveinhardt, "Trade in Central and Eastern European Countries Ten Years After Their EU Accession – Is There Convergence?" *Society and Economy* 37, no. 4 (2015): 443–60.
17 Simon Tilford, "Poland and Ukraine: A Tale of Two Economies," CER *Bulletin* 95 (April/May 2014): 2–3.
18 Council of the European Union, *Joint Declaration of the Eastern Partnership Summit*, Warsaw, September 29–30, 2011.
19 Council of the European Union, *Joint Declaration.*
20 Council of Europe, "Warsaw Declaration – Third Summit of Heads of State and Government of the Council of Europe," Warsaw, May 17, 2005.
21 Sergei Lavrov quoted Samual Charap and Timothy J. Cotton, *Everyone Loses: The Ukrainian Crisis and the Ruinous Contest for Post Soviet Eurasia* (London: ISSS, 2006), 100.
22 Alexander Cooley, "Whose Rules, Whose Sphere? Russian Governance and Influence in Post-Soviet States," Carnegie Endowment for International Peace, June 30, 2017.
23 Vladimir Putin, "Novyi Internatsionnyi Proekt Dlya Evrazii – Budushchee, Kotoroe Rozhdaetsya Sevodniya." *Izvestia*, 3 October 2011.
24 Andrew Osborn and Andrey Ostroukh, "Putin Rues Soviet Collapse as Demise of 'Historical Russia," *Reuters*, December 12, 2021.
25 Amanda Paul, "Russia Cranks Up Pressure on Ukraine," *Euobserver*, September 2, 2013.
26 Victor Zaborsky, "Crimea and the Black Sea Fleet in Russian-Ukrainian Relations," *CSIA Discussion Paper* 95-11, Kennedy School of Government, Harvard University, September 1995.
27 Reuters Staff, "TIMELINE: Russian cuts gas to Ukraine in latest gas row," *Reuters*, January 1, 2009.
28 Michael Kahn and Jan Lopatka, "EU-Russia gas duel deepens with Slovakia supply cut," *Reuters*, September 30, 2014.
29 Lawrence Freedman, *Ukraine and the Art of Strategy* (New York: Oxford University Press, 2019), 72.
30 Steven Pifer, "Why Did Ukraine's Yanukovych Give in to Russian Pressure on EU Deal?" Brookings, December 2, 2013.
31 Shaun Walker, "Vladimir Putin Offers Ukraine Financial Incentives to Stick with Russia," *The Guardian*, December 18, 2013.
32 Mykola Azarov cited in: Reuters Staff, "Ukraine says decision on EU is tactical, motivated by economics," Reuters, November 22, 2013.

33 Azarov cited in: Reuters Staff.

34 Freedman, *Ukraine and the Art of Strategy*, 74.

35 David Herszenhorn, "Ukraine Blames I.M.F. for Halt to Agreements with Europe," *New York Times*, November 22, 2013.

36 Christine Lagarde cited in: Kenneth Rapoza, "Russia Helped Ukraine, But Now Ukraine Needs More, IMF's Lagarde Says," *Forbes*, April 3, 2014.

37 Gunta Pastore, "The EU-Ukraine Association Agreement Prior to the Vilnius Eastern Partnership," *Baltic Journal of European Studies* 4, no. 2 (2017): 16.

38 Ambrose Evans-Pritchard, "Historic Defeat for EU as Ukraine Returns to Kremlin Control," *Telegraph*, November 22, 2013.

39 Walker, "Vladimir Putin Offers Ukraine Financial Incentives."

40 Serhiy Leshchenko, "Yanukovych's Secret Diaries."

41 BBC Newsnight, "Snipers at Maidan: The Untold Story of a Massacre in Ukraine – Newsnight," February 12, 2015.

42 Michael Kofman, Katya Migacheva, Brian Nichiporuk, Andrew Radin, Olesya Tkacheva, and Jenny Oberholtzer, *Lessons from Russia's Operations in Crimea and Eastern Ukraine* (Santa Monica, CA: RAND Corporation, 2017), 85.

43 Howard Colby and Ruslan Pukhov, *ed. Brothers Armed: Military Aspects of the Crisis in Ukraine* (Minneapolis, MN: Eastview Press, 2014).

44 Foreign Minister Sergey Lavrov's Remarks and Answers to Media Questions at the Primakov Readings International Forum, Moscow, June 30, 2017.

45 Lavrov's remarks and answers.

46 Gerard Toal, Near Abroad, Putin, the West, and the Contest over Ukraine and the Caucasus (Oxford: 2017; online edn, Oxford Academic, 12 November 2020), 199.

47 Commission on Security and Cooperation in Europe, "The December 1, 1991, Referendum/Presidential Election in Ukraine," CONGRESS, 2nd Session, 1992.

48 J.L. Black, *Russia & Eurasia Documents Annual 1997: The Russian Federation* 1 (Academic International Press, 1998), 129.

49 Luke Harding, "Ukraine extends lease for Russia's Black Sea Fleet," *The Guardian*, April 21, 2010.

50 "Novaya Gazeta's 'Kremlin Papers' Article," *UNIAN*, February 25, 2015.

51 Putin cited in: Andrey Kondrashov, "Crimea the Way Home," March 15, 2015.

52 Putin cited in Carl Schreck, "From 'Not Us' to 'Why Hide It?': How Russia Denied Its Crimea Invasion, Then Admitted It," *Radio Free Europe/Radio Liberty*, February 26, 2019.

53 Ibid.

54 Putin cited in: Kondrashov, "Crimea the Way Home."

55 Mark Galeotti, *Armies of Russia's War in Ukraine* (Oxford: Osprey Publishing, 2019), 11–12.

56 Howard Amos, "Ukraine: Sevastopol Installs Pro-Russian Mayor as Separatism Fears Grow," *The Guardian*, February 25, 2014.

57 Kent DeBennedictus, *Russian 'Hybrid Warfare' and the Annexation of Crimea: The Modern Application of Soviet Political Warfare* (London: Bloomsbury, 2022), 177–178.

58 Meduza, "Polite People: Who Are They?" *Special Report*, 2015.

59 DeBennedictus, *Russian 'Hybrid Warfare,'* 178.

60 Kostiantyn Kliatskin, "Crimea as It Was, Babylon," 13.

61 Meduza, "Polite People."

62 Colby and Pukhov, *Brothers Armed.*

63 Alexey Nikolsky, "Little, Green and Polite: The Creation of Russian Special Operations Forces," in *Brothers Armed: Military Aspects of the Crisis in Ukraine*, ed. Colby Howard and Ruslan Pukhov (Minneapolis: East View Press, 2015).

64 DeBennedictus, *Russian 'Hybrid Warfare,'* 179.

65 Yuri Petrovich Fedash cited in: DeBennedictus, *Russian 'Hybrid Warfare,'* 179.

66 Sergei Mironov cited in: Russia Must Protect Russians in Crimea," *Vesti*, February 26, 2014. As of July 28, 2016, Ramzan Kadyrov cited in: Russia Will Not Allow to Give Ukraine into the Hands of the Bandits [Рамзан Кадыров: Россия не позволит отдать Украину в руки бандитам]," *LifeNews*, February 26, 2014.
67 Ministry of Defense of the Russian Federation, "Forces Involved in the Verification of Combat Readiness Are Transferred to Designated Areas [in Russian]," February 27, 2014.
68 Kremlin, "Vladimir Putin Observed Western and Central Military District Forces Exercises," March 3, 2014.
69 Kofman et al., *Lessons from Russia's Operations*, 8.
70 NSDC, "Verbatim Report of the Meeting of the National Security and Defense Council of Ukraine under the Chairmanship of the Acting President of Ukraine," Chairman of the Verkhovna Rada of Ukraine O.V. Turchynov [in Ukrainian]," February 28, 2014.
71 DeBennedictus, *Russian 'Hybrid Warfare,'* 197.
72 Euromaidan Press, "Ukraine War: Russian Special Forces Seize Parliament Building in Crimea Ukraine," August 16, 2014.
73 A. Pavlushko, "How 'Ukrainian Berkut Officer' from Russian Ulyanovsk Assaulted Crimean Parliament Back in 2014," *Inform Napalm*, July 9, 2015.
74 Vladimir Chizhov cited in: Euronews, "Euronews Spoke to Russia's Ambassador to the EU Vladimir Chizhov," February 28, 2014.
75 Kremlin, "Vladimir Putin Answered Journalists' Questions on the Situation in Ukraine," March 4, 2014.
76 Harriet Salem, Shaun Walker, and Luke Harding, "Conflict fears rise after pro-Russian gunmen seize Crimean parliament," *The Guardian*, February 28, 2014; Shaun Walker, "Ukraine crisis: Crimea now part of Russia, local parliament declares," *The Guardian*, March 6, 2014.
77 Simon Shuster, "Putin's Man in Crimea Is Ukraine's Worst Nightmare," *Time*, March 10, 2014.
78 Kofman et al., *Lessons from Russia's Operations*, 8.
79 Radio Liberty, "Chronology of the Annexation of Crimea," *Euromaidan Press*, March 5, 2015.
80 Howard and Pukhov, *Brothers Armed*.
81 QHA, "State Broadcaster 'Crimea' Seized by Armed Men [in Russian]," February 28, 2014; UNIAN, "The State Television and Radio Company 'Crimea' Is Working under the Supervision of Armed Men [in Ukrainian]."
82 DeBennedictus, *Russian 'Hybrid Warfare,'* 180.
83 Sergei Loiko, "New Crimea Leaders Move up Referendum Date," *Los Angeles Times*, March 1, 2014.
84 Evgeniy Poddubny, "In Simferopol, Unknown Persons in Masks Attacked the Operator VGTRK [in Russian]," March 1, 2014; Stopfake.org, "Russian Blogger Exposed the Staging of a Shootout in Simferopol [in Russian]," March 3, 2014.
85 Maxim Popenker and N.R. Jenzen-Jones, "The Russian GM-94 Grenade Laucnher," ARES (2015).
86 Shaun Walker, "Ukraine Navy Officers Reject Plea to Defect to Russian-backed Crimea," *The Guardian*, March 3, 2014.
87 UN Security Council, "S/2014/146: Letter Dated 3 March 2014 from the Permanent Representative of the Russian Federation to the United Nations Addressed to the Secretary-General," March 3, 2014.
88 Kremlin, "Vladimir Putin Answered Journalists."
89 Verkhovna Rada of Crimea (2014c), No, 1702-6/14, "The Decree of the Verkhovna Rada of the Autonomous Republic of Crimea on Holding a General Referendum [in Russian]," March 6, 2014.
90 Verkhovna Rada of Crimea (2014c), "The Decree of the Verkhovna Rada."

91 Kremlin, "Vladimir Putin Answered Journalists."
92 Kofman et al., *Lessons from Russia's Operations*, 10.
93 Shane Harris, "Hack Attack," Foreign Policy, March 3, 2014.
94 "How the Audience of Ukrainian TV Channels Changed in Crimea [Як змінився перегляд українських телеканалів у Криму]," *Forbes*, April 2, 2014. As of November 6, 2015.
95 Roger McDermott, *Brothers Disunited: Russia's Use of Military Power in Ukraine* (Fort Leavenworth, Kan.: Foreign Military Studies Office, 2015).
96 I. Zhegulyov, A. Sivtsova, and J. Skibitskaya, "'No one Believed That This Was Serious': How Crimea Was Annexed: The Spring of 2014 Through the Eyes of Moscow, Kyiv, and Sevastopol [in Russian]," *Meduza*, March 21, 2014.
97 Golos, "The Ballot Form for Voting at the General Referendum on March 16, 2014 [in Russian]," *Twitter*, March 7, 2014.
98 Verkhovna Rada of Ukraine, "No. 350-XIV, On the approval of the Constitution of the Autonomous Republic of Crimea" [in Russian], December 23, 1998.
99 Mikhail Malyshev, "The Results of the Referendum in Crimea Read by Mikhail Malyshev [online video in Russian]," *112 Ukraina*, March 17, 2014.
100 Sevastopol City Council, "The Session of the City Council Approved the Results of the General Referendum on March 16, 2014 [in Russian]," March 17, 2014.
101 Kremlin, "Address by President of the Russian Federation," March 18, 2014.
102 BBC, "Crimea Referendum: Voters "Back Russia Union"," March 16, 2014; David Herszenhorn, "Crimea Votes to Secede from Ukraine as Russian Troops Keep Watch," *The New York Times*, March 16, 2014; Luke Harding and Shaun Walker, "Crimea Votes to Secede from Ukraine in "Illegal" Poll," *The Guardian*, March 16, 2014; Adam Withnall, "Crimeans Overwhelmingly Vote to Leave Ukraine and Join Russia in Contentious Referendum," *Independent*, March 16, 2014.
103 Verkhovna Rada of Crimea, "No, 1745-6/14, Resolution of the State Council of the Republic of Crimea on the Independence of Crimea [in Russian], March 17, 2014.
104 Kremlin, "Executive Order on Recognising Republic of Crimea," March 17, 2014.
105 CBS News, "Ukraine Troops Leave Crimea by Busload; Defense Minister Resigns After Russia Seizes Peninsula," March 25, 2014; bmpd (user), "The Defection of Ukrainian Navy Troops into the Russian Armed Forces After Leaving Crimea [Переход военнослужащих ВМС Украины в Вооруженные Силы России после оставления Крыма]," *Livejournal blog*, March 5, 2016.
106 Putin cited in: Kondrashov, "Crimea the Way Home."
107 Putin cited in: Kondrashov, "Crimea the Way Home."
108 "Transcript of a Secret Meeting of the National Security Council February 28, 2014 [Стенограмма секретного заседания СНБО 28 февраля 2014 года]," *ECHOMSK blog*, February 23, 2016. As of July 28, 2016.
109 Transcript of a secret meeting.
110 Justin Huggler, "Putin Privately Threatened to Invade Poland, Romania, and the Baltic States," *Daily Telegraph*, September 19, 2014.
111 Amy Knight, "Obama's Putin Delusion," *New York Review of Books* (blog), March 29, 2014.
112 Marc Thiessen, "Obama's Weakness Emboldens Putin," *The Washington Post*, March 3, 2014.
113 Nicholas Watt, "Ukraine: UK to Push for Tougher Sanctions against Russia over Crimea," *The Guardian*, March 18, 2014.
114 United Nations, "General Assembly Adopts Resolution Calling upon States Not to Recognize Changes in Status of Crimea Region," General Assembly, March 27, 2014.
115 Mariana Budjeryn, "Ukraine and the Treaty on the Non-Proliferation of Nuclear Weapons," *Wilson Center*, October 15, 2018.
116 Sam Jones, "Masterly Russian Operations in Ukraine Leave NATO One Step Behind," *Financial Times*, June 8, 2014.

Bibliography

Amos, Howard. "Ukraine: Sevastopol Installs Pro-Russian Mayor as Separatism Fears Grow," *The Guardian*, February 25, 2014, https://www.theguardian.com/world/2014/feb/25/ukraine-sevastopol-installs-pro-russian-mayor.

Barry, Ellen. "Former Ukraine Premier Is Jailed for 7 Years," *The New York Times*, October 11, 2011, https://www.nytimes.com/2011/10/12/world/europe/yulia-tymoshenko-sentenced-to-seven-years-in-prison.html.

BBC. "Crimea Referendum: Voters "Back Russia Union"," March 16, 2014, https://www.bbc.com/news/world-europe-26606097.

BBC Newsnight. "Snipers at Maidan: The untold story of a massacre in Ukraine – Newsnight," February 12, 2015, https://www.youtube.com/watch?v=mJhJ6hks0Jg.

Bivings, Lili. "Ukraine's Orange Revolution," *The Kyiv Independent*, August 2, 2023, https://kyivindependent.com/ukraines-orange-revolution/.

Black, J.L. *Russia & Eurasia Documents Annual 1997: The Russian Federation 1* (Academic International Press, 1998).

BMPD (user), "The Defection of Ukrainian Navy Troops Into the Russian Armed Forces After Leaving Crimea [Переход военнослужащих ВМС Украины в Вооруженные Силы России после оставления Крыма]," Livejournal blog, March 5, 2016.

Budjeryn, Mariana. "Ukraine and the Treaty on the Non-Proliferation of Nuclear Weapons," *Wilson Center*, October 15, 2018, https://www.wilsoncenter.org/blog-post/ukraine-and-the-treaty-the-non-proliferation-nuclear-weapons.

Caryl, Christian. "Yanukovych Won. Get Over It," *Foreign Policy*, February 8, 2010, https://foreignpolicy.com/2010/02/08/yanukovych-won-get-over-it/.

CBS News. "Ukraine troops leave Crimea by busload; defense minister resigns after Russia seizes peninsula," March 25, 2014, https://www.cbsnews.com/news/ukraine-troops-leave-crimea-by-busload-defense-minister-resigns-after-russia-seizes-peninsula/.

Charap, Samual, and Timothy J. Cotton. *Everyone Loses: The Ukrainian Crisis and the Ruinous Contest for Post Soviet Eurasia* (London: ISSS, 2006), 100.

Colby, Howard, and Ruslan Pukhov. eds., *Brothers Armed: Military Aspects of the Crisis in Ukraine* (Minneapolis, MN: Eastview Press, 2014).

Commission on Security and Cooperation in Europe. "The December 1, 1991 Referendum/Presidential Election in Ukraine," CONGRESS, 2nd Session, 1992, https://www.csce.gov/wp-content/uploads/2016/06/120191UkraineReferendum.pdf.

Cooley, Alexander. "Whose Rules, Whose Sphere? Russian Governance and Influence in Post-Soviet States," Carnegie Endowment for International Peace, June 30, 2017, https://carnegieendowment.org/posts/2017/06/whose-rules-whose-sphere-russian-governance-and-influence-in-post-soviet-states?lang=en.

Council of the European Union. *Joint Declaration of the Eastern Partnership Summit*, Warsaw, September 29-30, 2011.

Council of the European Union. "Warsaw Declaration - Third Summit of Heads of State and Government of the Council of Europe," Warsaw, May 17, 2005, https://www.consilium.europa.eu/media/31798/2011_eap_warsaw_summit_declaration_en.pdf

DeBennedictus, Kent. *Russian 'Hybrid Warfare' and the Annexation of Crimea: The Modern Application of Soviet Political Warfare* (London: Bloomsbury, 2022).

Euromaidan, PR. "Ukraine War: Russian Special Forces Seize Parliament Building in Crimea Ukraine," August 16, 2014.

Euronews. "Euronews spoke to Russia's ambassador to the EU Vladimir Chizhov," February 28, 2014, https://www.euronews.com/2014/02/28/euronews-spoke-to-russia-s-ambassador-to-the-eu-vladimir-chizhov.

Evans-Pritchard, Ambrose. "Historic Defeat for EU as Ukraine Returns to Kremlin Control," *Telegraph*, November 22, 2013, https://www.telegraph.co.uk/finance/comment/ambroseevans_pritchard/10469065/Historic-defeat-for-EU-as-Ukraine-returns-to-Kremlin-control.html.

Finn, Peter. "Widespread Vote Fraud Is Alleged In Ukraine," *The Washington Post*, November 23, 2004, https://www.washingtonpost.com/archive/politics/2004/11/23/widespread-vote-fraud-is-alleged-in-ukraine/e4c76edb-7e7f-4e9c-83f6-fe2fe321eb28/.

Finn, Peter. "Ukrainian President Turns to A Rival," *The Washington Post*, September 23, 2005, https://www.washingtonpost.com/archive/politics/2005/09/23/ukrainian-president-turns-to-a-rival/32708016-4cdc-412e-aeb0-fa9b078d6342/.

Freedman, Lawrence. *Ukraine and the Art of Strategy* (New York: Oxford University Press, 2019).

Galeotti, Mark. *Armies of Russia's War in Ukraine* (Oxford: Osprey Publishing, 2019).

Golos. "The Ballot Form for Voting at the General Referendum on March 16, 2014 [in Russian]," Twitter, March 7, 2014, https://twitter.com/golosinfo/status/441935863967399936/photo/1.

Grančay, Martin, Ērika Šumilo, and Jolita Vveinhardt. "Trade in Central and Eastern European Countries Ten Years after Their EU Accession - Is There Convergence?" *Society And Economy* 37, no. 4 (2015): 443–460, http://www.Jstor.Org/Stable/43774230.

Harding, Luke. "Ukraine extends lease for Russia's Black Sea Fleet," *The Guardian*, April 21, 2010, https://www.theguardian.com/world/2010/apr/21/ukraine-black-sea-fleet-russia.

Harris, Shane. "Hack Attack," *Foreign Policy*, March 3, 2014, https://foreignpolicy.com/2014/03/03/hack-attack/.

Herszenhorn, David. "Ukraine Blames I.M.F. for Halt to Agreements with Europe," *New York Times*, November 22, 2013, https://www.nytimes.com/2013/11/23/world/europe/ukraine-blames-imf-for-collapse-of-accord-with-european-union.html.

Herszenhorn, David. "Crimea Votes to Secede From Ukraine as Russian Troops Keep Watch," *The New York Times*, March 16, 2014, https://www.nytimes.com/2014/03/17/world/europe/crimea-ukraine-secession-vote-referendum.html.

How the Audience of Ukrainian TV Channels Changed in Crimea [Як змінився перегляд українських телеканалів у Криму]," *Forbes*, April 2, 2014. As of November 6, 2015: http://forbes.net.ua/ua/business/1368547-yak-zminivsya-pereglyad-ukrayinskih-telekanaliv-u-krimu.

Huggler, Justin. "Putin Privately Threatened to Invade Poland, Romania, and the Baltic States," *Daily Telegraph*, September 19, 2014, http://www.telegraph.co.uk/news/worldnews/europe/russia/11106195/Putin-privately-threatened-to-invade-Poland-Romania-and-the-Baltic-states.html.

Jones, Sam. "Masterly Russian Operations in Ukraine Leave NATO One Step Behind," *Financial Times*, June 8, 2014, https://www.ft.com/content/a5829d60-ed48-11e3-8a1e-00144feabdc0#axzz3QvHuCjQh.

Kadyrov. Ramzan cited in: Russia Will Not Allow to Give Ukraine into the Hands of the Bandits [Рамзан Кадыров: Россия не позволит отдать Украину в руки бандитам]," LifeNews, February 26, 2014. As of July 28, 2016: http://lifenews.ru/news/128000.

Kahn, Michael, and Jan Lopatka. "EU-Russia gas duel deepens with Slovakia supply cut," *Reuters*, September 30, 2014, https://www.reuters.com/article/uk-ukraine-crisis-gas-idUKKCN0HQ30720141001.

Kliatskin, Kostiantyn. "Crimea as It Was, Babylon'13, 2018, https://babylon13.org.ua/crimea-as-it-was/?lang=en.

Knight, Amy. "Obama's Putin Delusion," *New York Review of Books* (blog), March 29, 2014, https://www.nybooks.com/online/2014/03/29/obamas-putin-delusion/.

Kofman, Michael, Katya Migacheva, Brian Nichiporuk, Andrew Radin, Olesya Tkacheva, and Jenny Oberholtzer, Lessons from Russia's Operations in Crimea and Eastern Ukraine. Santa Monica, CA: RAND Corporation, 2017, 85, https://www.rand.org/pubs/research_reports/RR1498.html.

Kondrashov, Andrey. "Crimea the Way Home," March 15, 2015, https://www.youtube.com/watch?v=4xg6f5McZtM.

Kremlin. "Vladimir Putin Observed Western and Central Military District Forces Exercises," March 3, 2014, http://en.kremlin.ru/events/president/news/20362.

Kremlin. "Vladimir Putin Answered Journalists' Questions on the Situation in Ukraine," March 4, 2014, http://en.kremlin.ru/events/president/transcripts/20366.
Kremlin. "Executive Order on Recognising Republic of Crimea," March 17, 2014, http://en.kremlin.ru/events/president/news/20596.
Kremlin. "Address by President of the Russian Federation," March 18, 2014, http://en.kremlin.ru/events/president/news/20603.
Kucherov, Samuel. "The Future of the Soviet Collective Farm," *American Slavic and East European Review* 19, no. 2 (1960): 180–201.
Leshchenko, Serhiy. "Yanukovych's Secret Diaries," *Euromaidan Press*, March 12, 2014, https://euromaidanpress.com/2014/03/12/yanukovychs-secret-diaries/.
Loiko, Sergei. "New Crimea leaders move up referendum date," Los Angeles Times, March 1, 2014, https://www.latimes.com/world/worldnow/la-fg-wn-crimea-referendum-date-20140301-story.html.
Luke, Harding, and Shaun Walker. "Crimea Votes to Secede from Ukraine in "Illegal" Poll," *The Guardian*, March 16, 2014, https://www.theguardian.com/world/2014/mar/16/ukraine-russia-truce-crimea-referendum.
Malyshev, Mikhail, "The Results of the Referendum in Crimea Read by Mikhail Malyshev [online video in Russian]," *112 Ukraina*, March 17, 2014, https://www.youtube.com/watch?v=EvQ1r2sr9SE.
McDermott, Roger. Brothers Disunited: Russia's Use of Military Power in Ukraine, Fort Leavenworth, Kan.: Foreign Military Studies Office, 2015, https://www.academia.edu/11853815/Brothers_Disunited_Russia_s_Use_of_Military_Power_in_Ukraine.
Meduza. "Polite People: Who Are They?" *Special Report*, 2015, https://meduza.io/special/polite/who.
Medvedev, Dimitri. "Dmitry Medvedev congratulated Viktor Yanukovych on his victory," *The Kremlin*, February 15, 2010, http://en.kremlin.ru/events/president/news/6873.
Ministry of Defense of the Russian Federation, "Forces Involved in the Verification of Combat Readiness Are Transferred to Designated Areas [in Russian]," February 27, 2014.
Mironov. Sergei cited in: Russia Must Protect Russians in Crimea," Vesti, February 26, 2014. As of July 28, 2016: http://pics.vesti.ru/videos/show/vid/580541/cid/7/.
Nikolsky, Alexey. "Little, Green and Polite: The Creation of Russian Special Operations Forces," in *Brothers Armed: Military Aspects of the Crisis in Ukraine*, Colby Howard and Ruslan Pukhov, ed. (Minneapolis: East View Press, 2015).
'Novaya Gazeta's "Kremlin Papers" Article,' UNIAN, February 25, 2015, https://www.unian.info/politics/1048525-novaya-gazetas-kremlin-papers-article-full-text-in-english.html.
NSDC. "Verbatim Report of the Meeting of the National Security and Defense Council of Ukraine under the Chairmanship of the Acting President of Ukraine," Chairman of the Verkhovna Rada of Ukraine O.V. Turchynov [in Ukrainian]," February 28, 2014, http://www.rnbo.gov.ua/files/2016/stenogr.pdf.
Osborn, Andrew, and Andrey Ostroukh. "Putin rues Soviet collapse as demise of 'historical Russia," *Reuters*, December 12, 2021, https://www.reuters.com/world/europe/putin-rues-soviet-collapse-demise-historical-russia-2021-12-12/.
Paul, Amanda. "Russia Cranks Up Pressure on Ukraine," *Euobserver*, September 2, 2013.
Pastore, Gunta. "The EU-Ukraine Association Agreement Prior to the Vilnius Eastern Partnership," *Baltic Journal of European Studies* 4, no. 2 (2017): 16.
Pavlushko, A. "How "Ukrainian Berkut Officer" from Russian Ulyanovsk Assaulted Crimean Parliament Back in 2014," *Inform Napalm*, July 9, 2015, https://informnapalm.org/en/how-ukrainian-berkutofficer-from-russian-ulyanovsk-assaulted-crimean-parliament-back-in-2014/.
Pifer, Steven, "Why Did Ukraine's Yanukovych Give in to Russian Pressure on EU Deal?" *Brookings*, December 2, 2013, https://www.brookings.edu/articles/why-did-ukraines-yanukovych-give-in-to-russian-pressure-on-eu-deal/.
Plokhy, Serhii. *The Gates of Europe: A History of Ukraine* (New York: Basic Books, 2015).

Poddubny, Evgeniy. "In Simferopol, Unknown Persons in Masks Attacked the Operator VGTRK [in Russian], March 1, 2014, https://www.vesti.ru/doc.html?id=1336405#.
Popenker, Maxim, and N.R. Jenzen-Jones. "The Russian GM-94 Grenade Laucnher," ARES (2015), https://armamentresearch.com/wp-content/uploads/2015/09/The-Russian-GM-94-Grenade-Launcher.pdf.
Primakov Readings International Forum. Moscow, June 30, 2017, https://www.wtca.org/news/the-primakov-readings-international-forum-2017.
Putin, Vladimir. "Novyi internatsionnyi proekt dlya Evrazii – budushchee, kotoroe rozhdaetsya sevodniya," Izvestia, 3 October 2011, available at http://izvestia.ru/news/502761.
QHA. "State Broadcaster 'Crimea' Seized by Armed Men [in Russian], February 28, 2014, https://rusaggression.gov.ua/en/interests-of-the-state-broadcaster-cfbdb08d244f54bd-276c03442ea1a539.html.
Radio Liberty. "Chronology of the Annexation of Crimea," *Euromaidan Press*, March 5, 2015, https://euromaidanpress.com/2015/03/05/chronology-of-the-annexation-of-crimea/.
Rapoza, Kenneth. "Russia Helped Ukraine, But Now Ukraine Needs More, IMF's Lagarde Says," *Forbes*, April 3, 2014, https://www.forbes.com/sites/kenrapoza/2014/04/03/russia-helped-ukraine-but-now-ukraine-needs-more-imfs-lagarde-says/?sh=3d02a9885487.
Reuters Staff. "TIMELINE: Russian Cuts Gas to Ukraine in Latest Gas Row," *Reuters*, January 1, 2009, https://www.reuters.com/article/us-russia-ukraine-gas-timeline-idUSTRE5000X420090102.
Reuters Staff. "Ukraine Says Decision on EU is Tactical, Motivated by Economics," *Reuters*, November 22, 2013, https://www.reuters.com/article/cbusiness-us-ukraine-eu-government-idCABRE9AL0AA20131122.
Salem, Harriet, Shaun Walker, and Luke Harding. "Conflict fears rise after pro-Russian gunmen seize Crimean parliament," *The Guardian*, February 28, 2014, https://www.theguardian.com/world/2014/feb/24/ukraine-crimea-russia-secession.
Schreck, Carl. "From 'Not Us' To 'Why Hide It?': How Russia Denied Its Crimea Invasion, Then Admitted It," *Radio Free Europe/Radio Liberty*, February 26, 2019, https://www.rferl.org/a/from-not-us-to-why-hide-it-how-russia-denied-its-crimea-invasion-then-admitted-it/29791806.html.
Sevastopol City Council. "The Session of the City Council Approved the Results of the General Referendum on March 16, 2014 [in Russian]," March 17, 2014, https://web.archive.org/web/20140722133147.
Shuster, Simon. "Putin's Man in Crimea Is Ukraine's Worst Nightmare," *Time*, March 10, 2014, https://time.com/19097/putin-crimea-russia-ukraine-aksyonov/.
Stopfake.org. "Russian Blogger Exposed the Staging of a Shootout in Simferopol [in Russian], March 3, 2014, http://www.stopfake.org/rossijskij-bloger-razoblachil-instsenirovku-perestrelki-v-simferopole/.
The Associated Press, "Ukrainian President Fires His Government," *The New York Times*, September 8, 2005, https://www.nytimes.com/2005/09/08/international/europe/ukrainian-president-fires-his-government.html.
The Avalon Project, "Foreign Relations of the United States: 1918 The Conclusion of the Peace of Brest Litovsk the Peace Treaty of Brest-Litovsk 1918," Yale Law School, 2008, https://avalon.law.yale.edu/20th_century/brest.asp.
Thiessen, Marc. "Obama's Weakness Emboldens Putin," *The Washington Post*, March 3, 2014, https://www.washingtonpost.com/opinions/marc-thiessen-obamas-weakness-emboldens-putin/2014/03/03/28def926-a2e2-11e3-84d4-e59b1709222c_story.html.
Tilford, Simon. "Poland and Ukraine: A Tale of Two Economies," *CER Bulletin* 95(April/May 2014): 2–3.
Toal, Gerard. *Near Abroad, Putin, the West, and the Contest Over Ukraine and the Caucasus* (Oxford: 2017; online edn, Oxford Academic, 12 Nov. 2020), 199, https://doi.org/10.1093/oso/9780190253301.001.0001.
"Transcript of a Secret Meeting of the National Security Council February 28, 2014 [Стенограмма секретного заседания СНБО 28 февраля 2014 года]," ECHOMSK

blog, February 23, 2016. As of July 28, 2016, https://gordonua.com/publications/stenogramma sekretnogo-zasedaniya-snbo-vo-vremya-anneksii-kryma-v-2014-godu-polnyy-tekst-na russkom-yazyke-121122.html.

Transparency International. "Corruptions Perceptions Index," https://www.transparency.org/en/cpi/2022.

Treisman, Daniel. "Why Putin Took Crimea: The Gambler in the Kremlin," *Foreign Affairs* 95, no. 3 (May/June 2016): 47–55, https://www.foreignaffairs.com/articles/ukraine/2016-04-18/why-putin-took-crimea.

UNIAN. The State Television and Radio Company 'Crimea' Is Working under the Supervision of Armed Men [in Ukrainian], https://www.unian.ua/politics/891322-derjavna-teleradiokompaniya-krim-pratsyue-pid-naglyadom-ozbroenih-lyudey.html.

United Nations. "General Assembly Adopts Resolution Calling upon States Not to Recognize Changes in Status of Crimea Region," General Assembly, March 27, 2014, https://press.un.org/en/2014/ga11493.doc.htm.

United Nations Security Council. "S/2014/146: Letter Dated 3 March 2014 from the Permanent Representative of the Russian Federation to the United Nations Addressed to the Secretary-General," March 3, 2014, https://digitallibrary.un.org/record/766526?ln=en. UN

Verkhovna Rada of Crimea. No, 1702-6/14, "The Decree of the Verkhovna Rada of the Autonomous Republic of Crimea [in Russian]," March 6, 2014a, http://crimea.gov.ru/act/11607.

Verkhovna Rada of Crimea. "No, 1745-6/14, Resolution of the State Council of the Republic of Crimea on the Independence of Crimea [in Russian], March 17, 2014b, http://crimea.gov.ru/act/11748.

Verkhovna Rada Ukraine. No, 1427-XII, Act of Declaration of Independence of Ukraine," August 24, 1991, https://static.rada.gov.ua/site/postanova_eng/Rres_Declaration_Independence_rev12.htm.

Walker, Shaun. "Vladimir Putin Offers Ukraine Financial Incentives to Stick with Russia," *The Guardian*, December 18 2013, https://www.theguardian.com/world/2013/dec/17/ukraine-russia-leaders-talks-kremlin-loan-deal.

Walker, Shaun. "Ukraine Navy Officers Reject Plea to Defect to Russian-backed Crimea," *The Guardian*, March 3, 2014, https://www.theguardian.com/world/2014/mar/03/ukraine-navy-officers-defect-russian-crimea-berezovsky.

Walker, Shaun. "Ukraine Crisis: Crimea Now Part of Russia, Local Parliament Declares," *The Guardian*, March 6, 2014, https://www.theguardian.com/world/2014/mar/06/ukraine-crisis-crimea-part-of-russia-local-parliament-declares.

Watt, Nicholas. "Ukraine: UK to Push for Tougher Sanctions against Russia over Crimea," *The Guardian*, March 18, 2014, https://www.theguardian.com/world/2014/mar/18/ukraine-uk-push-sanctions-russia-crimea.

Wilson Center. "Orange Revolution: Ukraine Votes for Change," Kennan Institute, March 8, 2005, https://www.wilsoncenter.org/article/orange-revolution-ukraine-votes-for-change.

Withnall, Adam. "Crimeans Overwhelmingly Vote to Leave Ukraine and Join Russia in Contentious Referendum," *Independent*, March 16, 2014, https://www.independent.co.uk/news/world/europe/crimea-referendum-how-why-and-where-next-for-soontobe-divided-ukraine-9195310.html.

Yuhas, Alan, and Raya Jalabi. "Crimea's Referendum to Leave Ukraine: How Did We Get Here?" *The Guardian*, March 13, 2014, https://www.theguardian.com/world/2014/mar/13/crimea-referendum-explainer-ukraine-russia.

Zaborsky, Victor. "*Crimea and the Black Sea Fleet in Russian-Ukrainian Relations*," CSIA Discussion Paper 95-11, Kennedy School of Government, Harvard University, September 1995, https://www.belfercenter.org/publication/crimea-and-black-sea-fleet-russian-ukrainian-relations.

Zhegulyov, I., A. Sivtsova, and J. Skibitskaya, ""No one Believed That This Was Serious": How Crimea Was Annexed: The Spring of 2014 through the Eyes of Moscow, Kyiv, and Sevastopol [in Russian]," *Meduza*, March 21, 2014, https://meduza.io/feature/2017/03/21/nikto-ne-veril-chto-eto-vseriez.

7 Iran's Proxy War in Lebanon

Introduction

The previous chapter discussed the Russian annexation of Crimea, during which the Russians achieved their political objective by using local state and nonstate partners rather than through unilateral action. This concept of operation had specific relevance to the political and security environment in Ukraine, as it provided Russian policymakers with an option between economic coercion and overt military intervention.

In this chapter, the focus will shift to Iran's proxy war in Lebanon, during which indirect special operations played a different, but again critical, role. The effects of the operation discussed in this chapter were not obtained in a directly foreseeable time frame but required a rather unforeseeable duration. Moreover, they are less visible in concrete terms. This case study is a perfect example of the broadness of special operations, covering traditional military activities and deeply penetrating various elements of civilian life. The campaign, conducted by Iran's Islamic Revolutionary Guard Corps (IRGC), shows how to successfully employ all instruments of national power (Diplomatic, Information, Military, Economic, or DIME) synergistically to achieve strategic effects.

Background

In January 1978, the Iranian population started to demonstrate against the autocratic regime of Mohammad Reza Pahlavi, then the country's ruling Shah. Over the next few weeks, tensions increased as massive strikes and heated demonstrations became violent. A revolution was in the making. The uprising included groups from a variety of political organizations and religious persuasions.[1] Although there were significant differences between these groups, their shared antipathy to the Shah and a strong desire to be free from foreign interference united them in their resistance.

The fundamental drivers behind the revolution, however, were the ordinary Iranian people, fed up with the spendthrift despotism of the Shah. They served at the frontline of the increasingly intense and violent demonstrations. While these signs of escalation became apparent, the Shah only announced minor reforms to

DOI: 10.4324/9781003365112-10

calm the protesters. For example, he released thousands of political prisoners in Tehran. Rather than de-escalating the situation, his actions further boosted the protesters' belief in their ability to overthrow him.[2] The protests started spreading across the country and became even more violent. On August 19, Islamist activists set fire to a crowded cinema in the city of Abadan, near the Persian Gulf, to protest Western influence. As a result, 480 civilians died, trapped in the burning building.[3] Because of the continuous escalation of violence, the government declared martial law on September 8. That same night, security forces fired directly on a group of activists demonstrating at Tehran's Jaleh Square. The incident that led to the death of at least twelve activists became known as Black Friday and motivated the Iranian people to continue their uprising against the Shah.

Abroad, Ayatollah Ruhollah Khomeini started to play a crucial role in the development of the revolution. He met daily with international media and discussed quotes and reports that further fueled the demonstrations. In his talks, he ensured that his message would strengthen unity among the demonstrators while avoiding expressing his personal objectives. When leaders of the various political parties visited him in Paris, Khomeini, who demonstrated the required authority as a senior Shiite cleric, became the face of Iranian opposition and the potential new leader of Iran.

On January 16, 1979, the revolution had become too powerful, which made the Shah flee the country. The Iranian people did not consider the civil government that remained behind as credible. At the same time, a military *coup d'état* was looming. The United States (US) tried to prevent such a coup by persuading Iran's military leaders to give the civil government time to prove itself. The arrival of Khomeini in Tehran on February 1 indicated that the end of the revolution was nearing. The Iranian people viewed him as the revolution's leader, while his closest supporters considered him their religious authority and new political leader.

Although Khomeini's return to Iran signaled the end of the revolution, resistance from the military remained. However, the defection of eight hundred military personnel from the Air Force to the rebels proved to be the turning point. After publicly supporting Khomeini, the defectors got into a firefight with the Imperial Guards, who were still loyal to the Shah. After the defectors received additional support from fighters belonging to Mujahedin-e Khalq (MEK), they were able to seize the Dowshan Tappeh airbase in Tehran. They then took all the weapons from the armory and distributed them through mosques and community centers to other fighters in the revolution.[4]

The takeover of the airbase inspired other revolutionaries who started to raid police stations throughout the country.[5] The weapons and ammunition obtained during these raids significantly strengthened the revolutionaries while making the military loyal to the Shah collapse. Massive defections started to occur, and when the commanders of the Imperial Guards announced their surrender, the Supreme Council of the Armed Forces quickly assembled on February 11, 1979, and drafted a declaration of independence.[6] The release of the declaration on the state radio later that day officially ended the era of the Iranian monarchy. The revolution was successful.

Members of the IRGC were among Khomeini's strongest supporters and considered themselves the revolution's defenders. The organization served as an instrument of hard power to defend and advance Khomeini's political ambitions. In the aftermath of the revolution, the IRGC's primary mission was to prevent political rivals from undermining Khomeini's position through intimidation and aggression. On several occasions, the IRGC had to put down uprisings against the Ayatollah, including in its home province of Azerbaijan.[7] The importance of the IRGC proliferated as Khomeini distrusted the traditional military (the *Artesh*). That organization had been responsible for Iran's security during the Shah's regime and had received substantial military support from the US Military Assistance Advisory Group (MAAG).[8] As a result, the Ayatollah feared a *coup d'état* and needed to marginalize the *Artesh* while strengthening the security provided by his trusted aids of the IRGC.

The war with Iraq, during which Iran suffered a horrific number of casualties, coupled with the inability of Iran's regular army to sustain an effective defense against the superior Iraqi military, was another important reason for the growing importance of the IRGC and the employment of irregular tactics. On September 22, 1980, Saddam Hussein invaded Iran, which started a protracted war. By invading Iran, Hussein wanted to prevent the revolution that had brought Khomeini to power from spilling over into his own country. Additionally, the Iraqi leadership considered a country with an absolute Shia majority a significant threat, as it could exploit existing sectarian tensions between Iraq's own Shia majority and Saddam Hussein's Sunni-dominated Baath party that was leading the Iraqi government. The protracted conflict, during which Iraq received substantial military support from the US, ended on August 8, 1988, and led to the death of nearly half a million Iranian troops.[9] These heavy losses shocked the country with a relatively young population.[10] At the same time, the Iranians learned from the Soviet–Afghan War (1979-1989) about the successes of the Afghan Mujahedeen, who, supported by the US and Saudi Arabia, drove the Soviet Union out of Afghanistan. The Afghan victory was mainly the result of its effective irregular warfare strategy and inspired the Iranians to defend and advance their national interests at low costs. Quickly and effectively, the IRGC developed itself into a comparable guerrilla force that fought a protracted irregular warfare campaign against the Iraqi military.

The IRGC described this way of warfare as follows:

> Scattered, tactical guerilla attacks and harassing operations opened a new phase in the war which not only forced Iraq to halt its advance, but enabled us to obtain precious experience at the cost of the lives of our martyrs, in spite of all negative propaganda, the significance of IRGC men as well as the value and importance of the popular forces began to dawn upon the world.[11]

The change in strategy indicated that the government recognized that given its limited resources, it had to pursue such an irregular approach (i.e., deploying the IRGC as a guerilla force) to achieve victory. The approach became known as "mosaic defense," centered around decentralized guerilla warfare, including small-scale raids and psychological warfare to influence and demoralize an opponent over an extended period.[12] The purpose of decentralization is to create redundancy in command-and-control.[13] To this end, IRGC units operated independently and applied swarming tactics.[14] The approach aims to quickly mobilize resistance fighters and empower local leaders in strategically important districts. The IRGC's irregular approach is thus inherently scalable and flexible.

From 2019, Major General Hossein Salami commands the IRGC, composed of an estimated 190,000 troops that perform a variety of missions.[15] These missions range from homeland defense to protecting Iran's theocratic system, including exporting the Islamic Revolution. The latter has provided the IRGC's leadership with significant influence in all instruments of national power (i.e., diplomatic, information, military, economic, financial, intelligence, and law enforcement, or DIME-FIL). The IRGC is thus an organization that transcends the traditional military instrument of power. Its primary targets, however, lay in the military realm. Its military capabilities, therefore, include forces that operate in all military domains (i.e., air, land, sea, space, and cyberspace). The IRGC also has its own intelligence service and controls the *Basij*, which are militia fighters deployed to provide internal security and enforce control over civil society. The IRGC has thus become a grand strategic tool that extends Iran's regional footprint and ability to project power discretely and at low costs. This financial aspect deserves attention.

Iran's defense spending is significantly lower than that of its adversaries. Iran spent nearly 2.7 percent of its GDP on defense annually between 2000 and 2014.[16] Over the next three years, Iran increased its defense spending because of its growing military presence throughout the region.[17] In 2017, the Iranian government announced it would further increase defense spending to a minimum of 5 percent of GDP through 2021.[18] As of 2021, however, the Iranian defense budget seems to be 2.3 percent of GDP, a decrease of 0.4 percent compared to a decade ago.[19] Iran's adversaries spend significantly more on defense. The defense budget of Saudi Arabia grew from 7.24 percent of its GDP in 2011 to 13.49 percent of GDP in 2015.[20] In 2016, however, Saudi funding was believed to be 25 percent of GDP, equaling nearly USD 80 billion.[21] With that budget, Riyadh was the fourth-largest defense spender, preceded only by India, China, and the US[22] In 2021, Saudi funding was back at 6.6 percent of GDP, which is still more than double the Iranian budget.[23] In 2022, Iran's principal enemy, Israel, lowered its defense spending from 15.4 percent of GDP in 2008 to 11.8 percent of GDP in 2020.[24] A year later, the budget was increased to 12.2 percent of GDP and remained the same for 2022, which is more than five times higher than Iran's defense spending.

Such a vast delta between Iran's budget and its adversaries necessitates an irregular approach using militant clients. Although it is unlikely that the funding for these activities is included in the annual defense spending figures, running militant clients is a lot cheaper than investing in advanced conventional military capabilities.[25] Also, given the sanctions imposed on Iran, procuring advanced weapons systems on the international market has become extremely difficult. As a result, pursuing its political objectives through client states is the most feasible and equally desirable strategy for Iran.

Lebanon became the first country in which Iran would employ this strategy.

Lebanon

Like many contemporary countries in the Middle East, the territory today known as Lebanon had been contested by the former Ottoman Empire and Western colonial powers in the sixteenth century. Starting in 1516, Sultan Selim I, then ruler of the Ottoman Empire, invaded Syria and Lebanon's territories. The Ottomans maintained control over the parts until the end of World War I, although their influence was limited to the urban areas. After the Great War ended in 1918, France made the predominantly Muslim Mount Lebanon Mutasarrifate region part of Greater Lebanon, which stood under French control. On September 1, 1926, France formed the Lebanese Republic with a democratically elected parliament. It took until November 26, 1943, before Lebanon became an official sovereign country. Figure 7.1 presents a map of modern sovereign Lebanon.

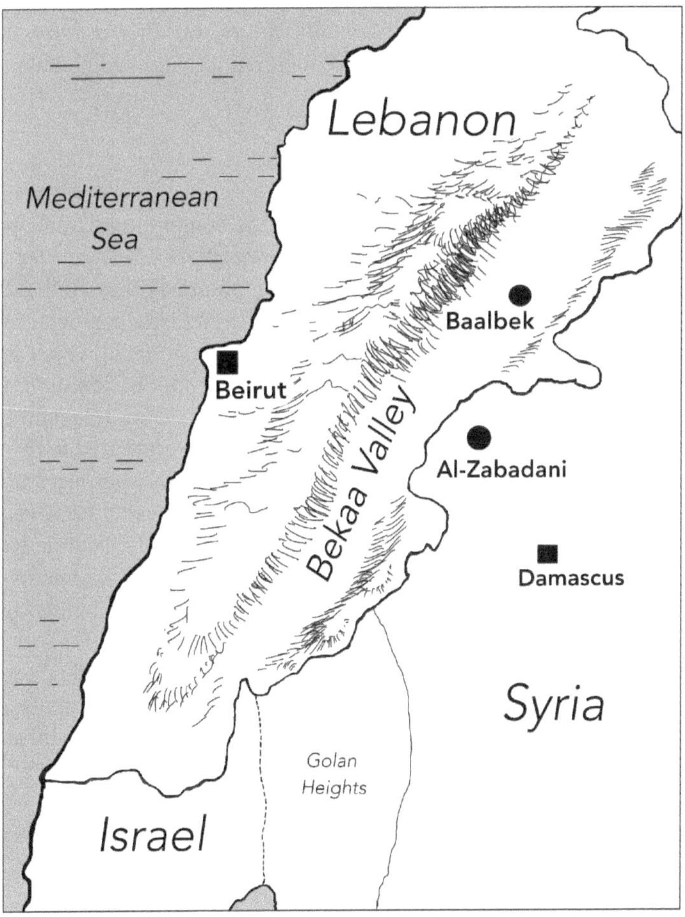

Figure 7.1 Map of Lebanon.

Since its independence, religious fault lines have significantly affected Lebanon's political stability. For example, the country's political system is characterized by confessionalism, reflected by the equal distribution of political power and influence among the Masonite Christians, Sunni Muslims, and Shia Muslims. The country's senior political leadership shows the significance of Lebanese sectarianism very well, as the president is always a Maronite Christian representative, the prime minister a Sunni Muslim, and the speaker of the parliament a Shiite Muslim.

Lebanon also had to cope with many Palestinians living in the country who were accused of forming a "state within a state." These tensions escalated in 1975 and led to a civil war devastating the country. Although the end of the civil war in 1990 provided an opportunity for some degree of peace and prosperity, a severe economic crisis struck the country in 2019 and pushed a substantial part of the population into poverty. Additionally, one and a half million Syrian refugees fleeing the Syrian civil war migrated to Lebanon. These dynamics illustrate that many of Lebanon's challenges have persisted into the twenty-first century, creating a volatile political and security situation and, thus, a population vulnerable to Iranian influence.

The IRGC and Hezbollah

The 1982 Israeli invasion of Lebanon signaled the beginning of Iran's proxy war against Israel. After a series of attacks and counterattacks between the Palestine Liberation Organization (PLO) and the Israeli Defense Forces (IDF), Israel decided to invade southern Lebanon on June 6, 1982. The Israelian objective was to force the PLO out of the indirect fire (i.e., rockets) range from which they could strike northern Israel. The invasion immediately drew a lot of attention, including from Iran. Khomeini and several of his closest advisers were reluctant to deliver a military response, as the Iranian military was heavily committed to the war with Iraq. Khomeini believed Iran had to deal with Lebanon only after the war with Iraq ended to prevent over-commitment.[26] Hardliners in the IRGC, however, favored military action, as they saw it as an opportunity to export the Islamic Revolution and fight their archrival, Israel.[27] After a quick internal debate, Khomeini deployed a small number of IRGC troops to support Lebanese and Palestinian fighters in their war against Israel.[28]

Subsequently, the IRGC announced that Iran would deploy military personnel to Lebanon to "engage in [a] face-to-face battle against Israel, the primary enemy of Islam and the Muslims."[29] Shortly after that, Khomeini himself publicly announced that the IRGC would deploy its troops to Lebanon:

> To us, there is no difference between the fronts in the south of Iran and in south Lebanon.... We are prepared to put our facilities and necessary training at the disposal of all the Muslims who are prepared to fight against the Zionist regime. We believe that victory will belong to the Muslims, and to those who are on the side of truth.[30]

That the preparation phase did not take long was shown by the fact that an advanced party of IRGC troops had already arrived in Damascus before Khomeini's announcement.[31] At that time, the Office of World Islamic Liberation Movements in Tehran was instrumental in deploying the IRGC operatives, then referred to as the "Lebanon Corps."[32] The office's mission was to establish networks between the IRGC and fellow Shia-Muslim organizations "fighting for freedom from the servitude and fetters of Western and Eastern imperialism and global Zionism."[33] Its focus was primarily on Lebanon and secondarily on Afghanistan.[34] A few years later, the office became an integral part of the IRGC, and the day-to-day management of operations became the responsibility of the IRGC's Intelligence Directorate.

Syrian President Hafez al-Assad, a Shia Muslim with strong ties with Iran, proved instrumental in the early phases of Iran's operations. For example, he allowed the IRGC to use Damascus International Airport for flying in troops and material destined for neighboring Lebanon. He also allowed IRGC operatives to move freely in Syrian territory. In that regard, the Syrian town of al-Zabadani, which lies just across the border with Lebanon, was significant as it served as the IRGC's logistic base and provided them with discrete access into the Bekaa Valley.[35] As a result, the IRGC was able to infiltrate more than 1,000 operatives into Lebanon in the first year of the campaign.[36] Additionally, the IRGC set up a discrete headquarters in a former Lebanese military base in Baalbek.[37] The town of Baalbek is situated in the Bekaa Valley, a Shia-dominated area of Lebanon, and thus, for the Iranians, a secure location from which to command-and-control their operations.

Iran's longstanding relationships with its Palestinian and Shiite partners in Lebanon played a vital role in the ease by which the IRGC could develop its operations.[38] The IRGC preferred to maintain close relationships with members of Shiite groups in the Bekaa Valley, whom they found to be more religious than the ones in Beirut.[39] Ali Akbar Mohtashemi-Pur, a senior Iranian cleric, was instrumental in the development of the Shiite network in Lebanon, particularly in the late 1970s, during which he frequently traveled to the Bekaa Valley while he served as Iran's Ambassador to Syria. During those years, coinciding with Lebanon's complex civil war, Mohtashemi built an effective coalition of Shiite militants and low-level clerics known as the pro-Khomeini Lebanese Shiite resistance. The movement ultimately coalesced into Hezbollah.[40]

Hezbollah became the first non-Iranian organization to conform to the Iranian revolution and accepted Khomeini as its supreme political, military, and religious authority.[41] Hezbollah's conformation to Khomeini's ideology gave Iran considerable influence over the organization. Several prominent and identifiable markers, such as Hezbollah's name (party of Allah) and official logo (based upon the IRGC's), illustrate the fundamental ties with Iran and their belief in the Velayat-e Faqih (Rule of the Jurisprudent).[42] The essence of the Velayat-e Faqih, rooted in Shia Islam, is the state's rule by the clergy – thus ultimately legitimizing the Ayatollah's rule. Notably, from its founding, Hezbollah consisted of a political party and a militia composed of its fighters.[43]

After the IRGC had laid the required fundament to support the Iranian long-term strategy, they started the next phase of the campaign: the development of political and military capabilities. Patience and a systematic step-by-step approach characterized the Iranian strategy from the start. From the beginning, IRGC operatives maintained low-profile and carefully developed human networks to collect intelligence, reconnoiter Israeli military positions, and shape the environment.[44] Recognizing the importance of gaining widespread support from the Lebanese Shia population, the IRGC also focused on establishing a social contract by providing them with essential services such as access to school, health care, and clean drinking water, among other things.[45] By doing so, the IRGC made its first yet vital steps toward gaining credibility, providing them with persistent access and freedom of movement in Shia-dominated areas.

The partnership with Hezbollah allows Iran to pursue its objectives through a layered network of organizations. The IRGC cooperates directly with Hezbollah, which collaborates with multiple other Shiite activist organizations on behalf of, or in line with, Iranian objectives. The IRGC builds its relationships on values and transactions. As clients, the organizations that the IRGC provides with advice, assistance, and additional support do not necessarily have to align with Iran's ideology entirely, nor do they take direct orders from the IRGC. Instead, the IRGC identifies and fosters organizations that seek complementary ends and have the will and capability to pursue them with Iranian support.[46] This active pursuit matters as much to defending and advancing Iranian national interests as achieving any client objectives in its neighboring states.[47]

Since this cooperation, the IRGC has consistently engaged in various activities to achieve its objectives within and outside Lebanese borders. These activities fall within four broad categories. The first category relates specifically to exporting the Islamic Revolution and includes propagating religious, cultural, and ideological norms, values, and beliefs. To this end, the IRGC distributes revolutionary literature among the Shiite population, including speeches of Khomeini and Ayatollah Ali Khameini (Khomeini's successor as Iran's Supreme Leader). The IRGC also brings Iranian religious scholars to Lebanon to teach courses in religion and revolutionary values.[48] In the second category, the IRGC provides military training to Shiite fighters. Apart from military training, time is devoted to ideology taught by Iranian clerics. The other two categories are related to Shia fighters' financial support and recruitment.[49]

In the early 1980s, the IRGC also supported Hezbollah in the planning, preparation, and funding of several high-profile attacks on Israeli command posts and Western Embassies as well as the multinational peacekeeping force (composed of France, Great Britain, Italy, and the US) that was deployed to Lebanon in August 1982 to support ending the armed conflict between the PLO, Syria (which supported the PLO), and Israel. The most notable attacks included the following suicide bombings:

- IDF Headquarters in Tyre (Lebanon) in November 1982, killing eighty-nine.
- US Embassy in Beirut on April 18, 1983, killing sixty-three people.

- Simultaneous car bomb attacks against the US and French barracks in Beirut on October 23, 1983, killing two hundred forty-one US Marines and fifty-eight French Paratroopers.
- Annex of the US Embassy in Beirut on September 20, 1984, killing twenty-four people.[50]

These attacks were conducted by the Islamic Jihad Organization (IJO), Hezbollah's terrorism wing. Headed by Imad Faiz Mughniyah, the IJO spearheaded the IRGC's most sensitive operations. As a former bodyguard of PLO leader Yasser Arafat, Mughniyah was recruited by the IRGC in Baalbek.[51] He proved well qualified for commanding clandestine operations, as an Israeli intelligence officer described: "He is one of the most creative and brilliant minds I have ever come across. He is a man with deep understanding, an excellent technical grasp, and leadership ability."[52] Mughniyah commanded the IJO until he was killed in a car bomb in Damascus in 2008, suspected to have been placed by the Mossad.

The attacks and their horrific high death toll of nearly five hundred people made US President Ronald Reagan decide to withdraw the US Marines from Lebanon in the spring of 1984 despite his expressed commitment to the peace process and holding those responsible accountable.[53] Due to uncertainty or complications of Cold War and Middle East policy, the US did not retaliate against Iran or Hezbollah, nor did the French. There was reportedly no US response against other Shiite militias either.[54] Hezbollah then picked up attacks against Israel and killed approximately 600 IDF soldiers between 1982 and 1984. As a result, Israel partially withdrew itself from Lebanon in 1985.[55] Over time, Hezbollah continued to conduct attacks against high-value targets, including political opponents. Most infamously, Hezbollah is accused of being involved in the assassination of former prime minister of Lebanon Rafic al-Hariri on February 14, 2005, in Beirut.[56]

A significant characteristic of the above attacks is the question of attribution. Contrary to most terrorist organizations, Hezbollah does not claim responsibility for attacks it is accused of. At the same time, the attribution process is complicated because a chain of actors and organizations is typically involved in the planning, preparation, and execution of an attack. For example, the Iranian government issues orders to the Office of World Islamic Liberation Movements, which passes the order and the requisite funding to the IRGC commander in Lebanon (note that the Iranian Embassy in Syria performed this function during the first years of the IRGC's operations). The IRGC then passes the order and allocates funding to Hezbollah, which uses its military wing ("Islamic Jihad") to execute the order. The chain of actors makes the attribution process slow, allowing both Hezbollah and Iran to plausibly deny any involvement.[57] Consequently, deterrence by punishment (i.e., delivering a timely and costly response) becomes ineffective.

An elite component of the IRGC, namely its Quds Force, serves at the forefront of these clandestine and covert operations. In March 1990, Khameini announced that the IRGC would deploy military consultants to advise and assist Iranian partners and allies. According to Khameini, "it is the legal obligation of the Revolutionary Guards to support liberation movements [i.e., to export the Islamic

Revolution]," and "arrangements have been made within the organization of the Revolutionary Guard to this effect."[58] The military consultants that Khameini referred to were members of the Quds Force (Quds means Jerusalem in Persian and Arabic) that emerged out of the IRGC's Lebanon Corps.[59]

From that moment on, the Quds Force would perform several of the IRGC's most discrete functions in Lebanon, including conducting strategic reconnaissance and training, funding, and arming Hezbollah. According to General (Ret.) Stanley McChrystal, the Quds Force, is best described as a combination of the Central Intelligence Agency (CIA) and the Joint Special Operations Command (JSOC).[60] General Qassem Soleimani served as its longtime commander until he was killed by an American drone strike on Baghdad International Airport on January 3, 2020. Soleimani played a vital role in the development of the Quds Force. According to Meir Dagan, former head of Israeli Mossad, Soleimani was a master in building networks as he had "a relationship with everyone," thereby contributing significantly to building strategic relationships with Iran's militant client states.[61]

Members of the Quds Force also became actively involved in offensive operations, including shaping the operational environment, conducting assassinations, and influencing the Lebanese population.[62] For example, after the IDF withdrew from the Lebanese security zone in May 2000, the Lebanese Armed Forces (LAF) did not want to replace them. Hezbollah did not hesitate and seized the opportunity by occupying the security zone instead. This action allowed Hezbollah to deploy its forces directly at the border with Israel. Together with Quds Force operatives and North Korean engineers, Hezbollah's Special Security Apparatus (a part of Hezbollah that cooperates closely with the Quds Force) began to build a secret and sophisticated network of tunnels and underground facilities.[63] Hezbollah used this underground interlinked system to perform critical warfighting functions such as command-and-control and logistics to support its combat operations against Israel. The facilities also offered protection and spaces to rest and provide medical care to Hezbollah fighters. Financial support provided by the Quds Force also facilitated the emplacement of a fiber-optic cable that ran from these underground facilities to Hezbollah's Headquarters in Beirut.[64] This allowed Hezbollah to operate secure communications during the 2006 Lebanon War.[65] Additionally, it allowed them to operate more discretely and unconventionally by using "shoot-and-scoot" tactics with their missile launchers. Tactics that Hezbollah learned from the IRGC.

Hezbollah's entrance into legitimate Lebanese politics is another example of Iran advancing its national interests. This political voice complements and supplements Hezbollah's militant operations. Hasan Nasrallah, Hezbollah's Secretary-General since 1992, explained that entry into Lebanese politics would increase support for Hezbollah's activities, provide access to and influence in any negotiations and parliamentary approved agreements, and improve representation and quality of life for the Lebanese Shia and lower classes in Lebanon's confessional political system. In September 1993, approximately one year after Hezbollah entered parliament, its representatives demanded an investigation into and punishment of LAF officers responsible for suppressing a Hezbollah-backed riot against the Israeli–Arab Peace Process.[66]

Next to its domestic activities, Hezbollah, while directed by Iran, employs its activities on a global scale. These activities are funded by a multi-million-dollar international drug trafficking business supported by a vast network of front companies and with access to Lebanese Shia émigré communities who assist Hezbollah with collecting intelligence, fundraising, providing logistics support, and even participating in conducting terrorist attacks. For example, the IRGC and Hezbollah have attacked the US, Israel, and Saudi Arabia across the Middle East, in South America, and even in the continental US They bombed the Israeli Embassy and another Israeli diplomatic facility in Buenos Aires in the early 1990s, killing one hundred and fifteen people. They are both implicated in the 1996 Khobar Towers military housing complex bombing in Saudi Arabia that killed nineteen and wounded five hundred and fifteen American military personnel.[67] In 2011, the US Drug Enforcement Agency (DEA) also disrupted an Iranian, likely IRGC, plot to assassinate the Saudi Ambassador in Washington, D.C.[68] All these activities illustrate how the IRGC could transform an organization composed of several loosely coupled Shiite militias fighting Israel into a powerful organization (i.e., Hezbollah) that has both political legitimacy and military capability, including an expeditionary one.[69]

Since 1997, the US has designated Hezbollah a foreign terrorist organization. The European Union (EU) followed in 2013 by designating the organization's armed wing as a terrorist organization. Its level of influence, however, extends far beyond any other terrorist organization in the world, both geographically and across elements of power. For example, when the situation in Iraq required an expansion of the Iranian training infrastructure, Quds Force commander Soleimani requested logistical support from Hezbollah to manage the training sites. The fact that Hezbollah's fighters speak Arabic, have a deep understanding of Arab culture, and bring videos to demonstrate to the trainees how they ambushed IDF forces increased the effectiveness of training the Shia militias in Iraq.[70] Hezbollah also played a leading role in preventing the fall of Syrian President Bashar al-Assad, notably during the major battles for Qalamoun, Aleppo, and Quasayr, among others. Losing al-Assad would significantly degrade Iranian influence in Syria and jeopardize Hezbollah's cooperation with the IRGC. Iran, therefore, persuaded Hezbollah to become actively involved in the war. Most notably, Hezbollah trained more than 250 Iraqi fighters to fabricate and use a new weapon, an explosively formed penetrator (EFP). Lucrative targets for the EFPs, which cost approximately USD 50 to produce, were American armored vehicles and M1A2 tanks. The results were dramatic, as the number of American and coalition military killed by EFPs between 2005 and 2011 was 609. Finally, Iran's disruption of the 2011 Status of Forces Agreement (SOFA) for US forces in Iraq shows Iran's ability to extend the Hezbollah model into more countries, and the Popular Mobilization Forces (PMF) growing influence shows further progress. Yemen's Zaydi Shiite Houthis offer an even more lucrative outcome for lower costs should the IRGC- and Hezbollah-backed rebels succeed in taking over the government. Iran could effectively envelop Saudi Arabia and threaten access to the Persian Gulf and the Red Sea through its clients.[71]

More than four decades after it started, the long-term cooperation between the IRGC and Hezbollah has significantly increased the Iranian political and military influence in Lebanon and the broader Middle East. Hezbollah's participation in the Lebanese government and direct provision of social services to the Lebanese people, including medical care during the COVID-19 pandemic, provides the organization and, by extension, Iran credibility and political influence in the country. Today, the Iranian influence in Lebanon is directly observable by the many Iranian flags and images of the Iranian Ayatollah and Soleimani in the Shia-dominated parts of Beirut, let alone the Bekaa Valley or southern Lebanon, and reflects Iran's popularity among the country's Shia population. Iran's partnership with Hezbollah extends its political and military influence beyond the Lebanese borders into Iraq, Syria, Yemen, and the Persian Gulf, further complicating international diplomacy and counterterrorism efforts in a complex and volatile region.[72]

Analysis

To determine the logic of the underlying Iranian strategy, the final part of this chapter analyzes whether (or to what degree) Iranian policy, strategy, and tactics align relative to the strategic, operational, and tactical levels of war.

Policy

The 1982 Israeli invasion of Lebanon signaled the beginning of Iran's proxy war against Israel. Khomeini responded immediately by sending the IRGC to assist its fellow Shia Muslims in their fight against the invaders. The protracted and bloody war with Iraq, which was ongoing at that time, did not prevent Khomeini from deploying the IRGC to advance his political objective: defeating the enemy. Some hardliners in the IRGC also saw the Israeli invasion as an opportunity to advance a second political objective: exporting the Islamic Revolution to Lebanon. Significantly, the American support provided to Iraq in its war with Iran further strengthened Iranian hardliners' belief that it had enemies at its borders and a great distance. Iran, therefore, wanted to decrease the influence of Western countries in the region, including Israel and the US After the partial Israeli withdrawal in 1985, the objectives became broader and included:

1 Maintaining power and influence over Hezbollah and, by extension, Lebanese politics.
2 Extrapolate the Hezbollah model to other countries in the Middle East to increase their presence and influence throughout the entire region while simultaneously decreasing Western influence.

Military strategy

Iranian leadership realized that the objectives could not be achieved swiftly but gradually over time and thus required a long-term strategy. In the early 1980s,

Iran's political leaders recognized that its limited budget and deteriorating relationships with the West would make it challenging to procure sophisticated military capabilities to conduct large-scale conventional warfare. Iran needed low-cost, high-return options to counter its relatively weak conventional military power and lack of nuclear deterrent to achieve its objectives. The preferred model became one of exporting violence and gaining influence through militant clients that would fight Iran's adversaries through sustained irregular warfare campaigns. The Iranians first developed such a strategy to achieve their objectives in Lebanon, in which Hezbollah played a vital role.

Iran developed a strategy to achieve four distinct objectives. First, the IRGC had to bring revolutionary literature, including speeches of Khomeini and other influential clerics, to Lebanon and distribute the materials among the Shia population. Additionally, Iranian religious scholars would travel to Lebanon to teach religious and revolutionary values courses. Through these actions, the Iranians could export the Islamic Revolution. Second, IRGC operatives would establish a base of operations in the Syrian town of al-Zabadani that would serve as their logistic hub. From there, the IRGC could discretely cross the border into the Lebanese Bekaa Valley that stood under the control of the Shia population. After a protracted preparation phase during which the IRGC operatives shaped the environment by collecting intelligence, building networks, and conducting reconnaissance, the IRGC would start recruiting and training the Shiite militants to launch attacks against high-value Israeli and Western targets. In addition to the military training, Iranian clerics joined the IRGC in Lebanon to teach the Shiite militants courses on ideology. Third, the IRGC would actively recruit Shiite fighters to join their efforts. Finally, Iran would provide financial support and management to sustain the campaign effectively.

Suitability: could Iran execute the operation with the available resources?

Iran successfully executed the operation and achieved all four objectives without significant challenges. In fact, the operation is still ongoing at the time of this writing. From the beginning of the operation, Iran assured it had kept enough IRGC operatives in Lebanon relative to the phase of the operation, thereby keeping the costs low. They also deployed enough religious clerics and teachers to export the ideology of Khomeini effectively. The relationship with Shiite militants was vital in establishing the social contract with the Lebanese Shia population during the operation's first and most critical phase, as the IRGC operatives, whose native language is Farsi, did not speak Arabic nor the local Levantine dialects spoken in Lebanon. Contrary to Mohtashemi, who frequently traveled to Lebanon, only a few IRGC operatives did so in the years before their employment. As a result, most operatives had limited knowledge of the country, its customs, culture, and complex clan structure. To succeed, Iran and the Shiite militants mutually depended on each other, demonstrating the value of the Iranian militant client strategy based on values and transactions.

Feasibility: did achieving the objectives create the strategic effects contributing to the desired end?

After the IRGC arrived in Lebanon in June 1982, it patiently prepared the operational environment. Once that was done, an estimated 1,000 IRGC troops started to recruit Shia fighters and run six training camps in the Bekaa Valley to train the Shia militants to fight the Israelis. These efforts resulted in the creation of a substantial number of Shia fighters capable of kidnapping American and European officials and conducting high-profile attacks against Israeli and Western targets throughout the country.[73] The first significant victory came after the bombings of the US Embassy and military bases of US and French forces in Beirut. The strategic effect of these attacks was that the US withdrew its troops from Lebanon. Following the US withdrawal, the IRGC and its Shia militia, now known as Hezbollah, continued to conduct attacks against Israel and taught them a painful lesson: Iran de facto borders Israel (i.e., via its proxy Hezbollah) while Israel does not border Iran. The training provided by the IRGC proved to be very effective and significantly improved Hezbollah's tactical performance on the battlefield. For example, in 1987, the ratio of Hezbollah and IDF soldiers that died in combat was 30-1, while this became 1-1 by 2000. As a result, Israel partially withdrew from Lebanon. In other words, the strategic effects (i.e., the withdrawal of US and Israeli forces) contributed to the Iranian objective of decreasing Israeli and Western influence in Lebanon.

In addition to the military activities, the IRGC provided salaries, medical care, and education to Hezbollah fighters and their families. This allowed the IRGC to establish a broader social contract with the Lebanese Shia population, which, in turn, provided the Iranians with the required legitimacy for their operations. The Shia population started to perceive Hezbollah and, by extension, Iran as genuine governing authorities. Importantly, from a moral and legal perspective, the Shia population perceived the actions of Iran and Hezbollah as righteous. Because of the broad support for Hezbollah among the Lebanese Shia population, it became a formal political party that secured its official position within the Lebanese government. Despite being designated as a terrorist organization by the US and EU, Hezbollah's official seat in Lebanese politics further increased Iran's power and influence over Lebanese politics.

After four decades of close cooperation, Hezbollah extends Iran's military and political influence into the Middle East, including Iraq, Syria, Yemen, and the Persian Gulf. For example, the US invasion of Iraq in 2003 and the subsequent overthrow of Saddam Hussein's Ba'ath Party, Iran's principal counterweight in the region, provided the Iranians with a strategic advantage and opened the door for Hezbollah to launch operations. In Syria, Hezbollah controls several critical border crossings and main supply routes leading to Damascus. Later in the Syrian civil war, Hezbollah became actively involved in combat operations supporting the al-Assad regime. Furthermore, Iran and Hezbollah actively support the Zaydi Shiite Houthis in Yemen's civil war. By exporting the Islamic Revolution to Yemen, Iran created a militant client close to its regional adversary, Saudi Arabia, that keeps them engaged in a protracted and costly war. These strategic effects contribute to achieving Iran's objective of extrapolating the Hezbollah model to other countries in the Middle East to increase its presence and influence throughout the entire region while simultaneously decreasing that of its adversaries.

Acceptability: did the strategic effects sought justify the objectives pursued and consider tangible and intangible costs?

Proxy warfare often limits escalation between sponsoring states by avoiding direct confrontation between the two and making attribution problematic. Iran exploits this by directing and supporting Hezbollah's attacks against Western military and intelligence personnel, diplomats, and civilians. Any direct Western response against Iran has been restricted to limited retaliations against specifically designated personnel or facilities approved at prominent levels of government. More often, the response was directed at facilities or militants associated with Iran inside the area of operations, thereby having only a limited effect.[74] This seems to indicate that Iran's strategy has proven to be effective.

However, Iran's sustained cooperation with Hezbollah in Lebanon and the broader region has also come with significant tangible and intangible costs. First, because the US and EU have designated the IRGC and Hezbollah as terrorist organizations, their leadership suffers from sanctions and travel restrictions. In addition, procuring modern weaponry and financing their operations remains exceedingly difficult, as the sanctions also target suspected IRGC front companies. Iran's strategy also poses the risk of increased Western counterterrorism operations against its militant clients and themselves. Over the last five decades, the Israeli or US military targeted Hezbollah and IRGC senior leaders. A notable example is the US drone strike that killed IRGC Quds Force commander Soleimani in 2020.

Second, Iran's operations created an increasingly hostile attitude among Sunni Muslims in the Middle East. Sunni-dominated countries consider Iran the greatest threat to regional peace and stability.[75] Furthermore, several of Iran's adversaries, including Israel, Saudi Arabia, and the United Arab Emirates (UAE), have taken a unified approach to counter Iran and are increasingly willing to commit their resources.[76] This coalition furthers the balance of conventional military forces to Iran's disadvantage.

Third, although actions from militant client states are difficult to attribute, several nations, including the US, have warned Iran that such actions could potentially lead to escalation.[77] For example, US retaliations were limited during the Obama Administration because of ongoing talks surrounding the nuclear deal. During the Trump Administration, threats against Iran became more explicit, particularly after Yemeni Houthis attacked US naval vessels. Next to additional sanctions imposed on the IRGC, the White House put Iran "on notice."[78] Iran's low-cost, high-return strategy would undoubtedly be at risk if the US or other powerful regional actors delivered a more robust response to their actions, including militarily.

Finally, Iran faces substantial political instability within its borders. Over the last decade, public dissent has led to mass demonstrations throughout the country. Iran's religious conservatism, which is also the driver behind its strategy of militant client states (i.e., exporting the Islamic Revolution), has been under pressure, particularly since the protests also occurred in rural areas, traditionally more aligned with the regime.[79] What stands out is the protesters' claim that the government spends more money on its militant clients than its citizens. Growing public dissent coupled with economic sanctions and other social vulnerabilities could potentially weaken the regime at home, which, in turn, could undermine Iran's position across the Middle East.

Risk

What follows is an assessment of the risks associated with Iran's proxy war in Lebanon.

What assumptions did the Iranians make in their strategy, and what was the effect of any wrong assumption?

Iran made several assumptions in its strategy that proved to be correct. For example, the Iranian belief that they could successfully export the Islamic Revolution to the Shia population of Lebanon allowed them to recruit, organize, train, and equip a substantial number of Shia fighters. These fighters initially formed pro-Iranian armed groups that later became Hezbollah. Iran also accurately assumed that the rural Shia population had little connection to the Lebanese government, which allowed Iran to establish a social contract with that population. By providing the local Shia population with basic goods and services such as health care and education, they perceived Hezbollah and the Iranians as credible and legitimate actors and even as supporters. Finally, Iran believed that employing a sustained irregular warfare campaign, including kidnapping US and European officials and striking high-profile targets, would decrease Western willingness to stay engaged in Lebanon. This assumption also proved correct, as the bombings on the US Embassy and American and French military barracks led to the withdrawal of US forces from the country. Iran then shifted its focus to Israel, which partially withdrew its forces less than a year later.

Did the Iranians incorporate flexibility or adaptability into their strategy? If required, how could they have modified their strategy, and at what cost?

The Iranian strategy was akin to running a marathon, not a sprint. Regarding strategy, the former needs flexibility over a longer duration, while the latter affords little time for adjustment. As early as the late 1970s, senior Iranian clerics traveled to Lebanon to establish networks and invest in relationships with the Shia population. When, more than a decade later, Israel invaded Lebanon, Iran did not deliver an immediate military response but instead deployed the IRGC to the country and slowly but surely created a powerful militant client organization. By doing so, Iran maintained a comprehensive view of the whole campaign while combining the employment of the various instruments of power relative to the condition of the strategic environment.

How did other actors react to the Iranian operation?

Given the protracted character of the Iranian operation, various actors, including local, regional, and international, have delivered favorable and unfavorable responses over time. Countries in the Middle East responded differently to the Iranian operations. Syria, for example, favored the Iranian activities in Lebanon, and al-Assad eased and concealed Iranian support to Hezbollah in its initial stages.[80] The IRGC has relied on Syrian support to access, train, and equip Hezbollah from the beginning of the operation. On the other hand, the operation has led Iran into an enduring Cold War with Saudi Arabia and deteriorated the relationships with other regional actors, including UAE, Bahrain, Egypt, Jordan, Kuwait, Pakistan, and Turkey.

Western countries condemned the Iranian operation and imposed sanctions against the front companies that finance the operation as well as the IRGC and its leadership. The US and EU designated the IRGC and Hezbollah (or its armed wing) as terrorist organizations. At the same time, Iran was able to deepen its relationships with the People's Republic of China (PRC) and the Russian Federation. Both major powers support Iran to ease the pressure of Western sanctions and yield material and logistical support for the IRGC and its proxies.[81] For example, Chinese companies are suspected of supplying Iranian proxies with unmanned aerial vehicles (UAVs) technology. Conversely, Iran provides UAVs to Russia in its ongoing war with Ukraine.[82]

How did unforeseen factors play a role in the strategy?

Iran executed its strategy without any unforeseen obstacles. Both scholars and practitioners have credited Iran with transforming Hezbollah from a loosely set of pro-Iranian fighters into a legitimate political actor in Lebanese politics, capable of conducting sustained military operations across the Middle East.[83] There are no indications that chance or friction played a role in this strategy.

Tactics

Did the Iranians achieve relative superiority throughout the campaign?

The IRGC achieved relative superiority by applying the principle of legitimacy and security. The Iranian proxy war in Lebanon is a clear example of how a quantitatively inferior force with less equipment than its opponents can generate strategic effects. The Iranian force sent into Lebanon in 1982 was small. It consisted of approximately 1,000 IRGC operatives, who were among Iran's best-trained troops and had gained substantial combat experience in the war with Iraq. This small force could offset the asymmetric advantage held by Israel and the US

The IRGC's tactic of proving its legitimacy to the Lebanese Shia population decreased the area of vulnerability. The IRGC did so by benefiting from the perspective of the Lebanese Shia population, who viewed their actions as legal, moral, and righteous, particularly since the IRGC was able to provide humanitarian and civil aid in cooperation with the Shia militias rather than the Lebanese government that

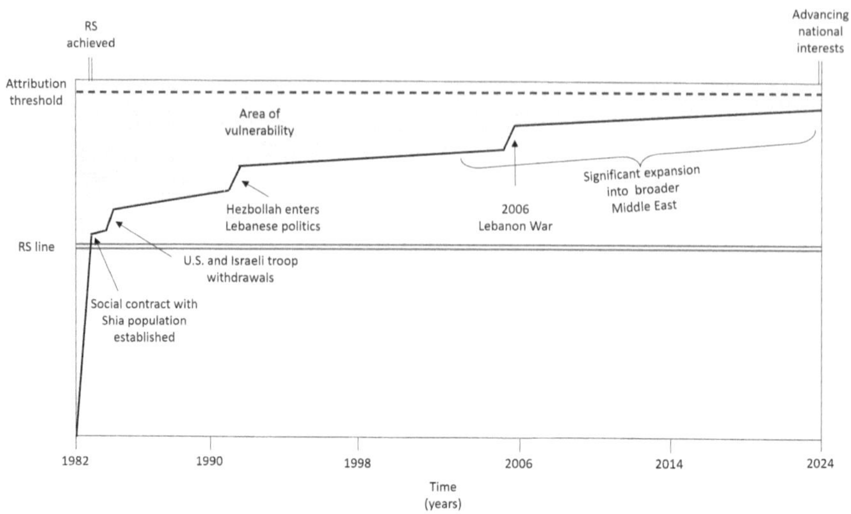

Figure 7.2 Relative superiority graph for Iran's proxy war in Lebanon.

is mostly absent in the Shia-dominated areas of the country. By heavily investing in the relationship with the local Shia population from the beginning of the campaign, the Iranians achieved relative superiority during the first years of the campaign.

Having achieved relative superiority at the beginning of the campaign meant that the IRGC and Hezbollah operatives had freedom of movement and action in the border area between Syria and Lebanon as well as in specific regions of the country, including the Bekaa Valley and the south. From now on, they could also rely on support from the local population. The combination of these advantages meant that the chances of mission success outweighed the chances of failure. Figure 7.2 illustrates this by the gradually inclining curve that reaches further above the Z-axis over time (i.e., relative superiority line), thereby decreasing the area of vulnerability. Of course, the operatives operating in the country remained vulnerable from the moment relative superiority was achieved. They mitigated that vulnerability by operating below the detection threshold that required strict measures of operational security.

If achieved, how did the Iranians sustain their relative superiority throughout the campaign? If not, why did they fail?

The IRGC sustained its relative superiority by applying the principle of perseverance like in a marathon. Establishing the social contract with the Shia population and developing Hezbollah into a significant force capable of striking the Israelis and American troops did not fully achieve the Iranian objectives. To further defend and advance Iran's objectives, Hezbollah had to become an official Lebanese political party. Also, Israel and the US had to withdraw their forces from the country, which required the Iranians to sustain their relative superiority. Iran realized that achieving its objectives decisively was complex and required a patient, committed, and persistent approach. By employing perseverance, the Iranians ensured the commitment necessary to achieve their objectives, including the withdrawal of Israeli and US forces from the country and the entrance of Hezbollah into Lebanese politics. Sustaining relative superiority was also achieved by applying strict measures of operational security. These strict measures made it impossible for foreign intelligence services to infiltrate the IRGC or Hezbollah, as their fighters proved to be immune to the four common pillars of spy recruitment: money, ideology, coercion, and ego.

Conclusion

Iran's proxy warfare in Lebanon illustrates that the chances of victory for weaker actors significantly increase when employing a different strategy than its opponent. Military theorists differentiate between direct strategies, such as conventional warfare, and indirect strategies, such as irregular warfare. Iran recognized that its limited conventional military power and lack of nuclear deterrent would lead to its defeat when fighting a conventional war with Israel and the US The alternative strategy, therefore, was to export violence and gain influence through militant clients that would contribute to defending and advancing Iran's national interest through sustained irregular warfare campaigns. By doing so, Iran could cultivate deep relationships with Hezbollah and, by extension, other networks.

The protracted operation illustrates that Iran's political leaders and military commanders have demonstrated a high level of strategic sense, as they well understood the conversion of tactical performance into strategic effect. The Iranian operation thus fulfilled the strategic promise of special operations by carefully aligning policy, strategy, and tactics.

Notes

1 Organizations participating in the revolution included the pro-democratic Liberation Movement of Iran, the National Front, the Islamist-Marxist Mojahedin-e-Khalq, the communist Tudeh Party, and the Marxists People's Fadai. Additionally, Islamists loyal to Ayatollah Khomeini played a significant role in the demonstrations as well.
2 Said Amir Arjomand, *The Turban for the Crown: The Islamic Revolution in Iran* (New York: Oxford University Press, 1988), 114 –119.
3 Afshon Ostovar, *Vanguard of the Imam, Religion, Politics, and Iran's Revolutionary guards* (New York: Oxford University Press, 2016).
4 Arjomand, *The Turban*, 124 –127.
5 Alireza Mahfoozi, in an interview recorded by Zia Shedghi, April 7, 1984. Paris, France. Iranian Oral History Collection, Harvard University, 19.
6 Sepehr Zabih, *The Iranian Military in Revolution and War* (New York: Routledge, 1988), 73.
7 Arjomand, *The Turban*, 139 –141.
8 Michael Knights and Michael Eisenstadt, "Mini-Hizballahs, Revolutionary Guard Knock-offs, and the Future of Iran's Militant Proxies in Iraq," *War on the Rocks*, May 9, 2017; Brigadier General Theo Mataxis, 1970, Senior Officer Debriefing Report: Chief Army Section, ARMISH/MAAG, Period April 1968 to May 1970, Department of the Army, Washington D.C., Confidential, July 16, declassified June 1980.

9 Dexter Filkins, "The Shadow Commander," *The New Yorker*, September 30, 2013.
10 Ibid.
11 A Glance at Two Years of War [Tehran?]: Islamic Revolution's Guards Corps [Political Office], 1982, 40–41.
12 J. Matthew McInnis, "Iran at War: Understanding Why and How Tehran Uses Military Force," December 2016, American Enterprise Institute.
13 John Arquilla and David Ronfeldt, *Swarming and the Future of Conflict* (Santa Monica, CA: RAND Corporation, 2000).
14 Anthony Cordesman, "Iran's Revolutionary Guards, the Al Quds Force, and Other Intelligence and Paramilitary Forces," August 16, 2007.
15 "Chapter Seven: Middle East and North Africa," *The Military Balance* 120, no. 1 (2020): 324–387.
16 SIPRI's data on Iran is accessible through the World Bank's open data website.
17 "Iran to Hike Military Spending Despite Lower Oil Prices, Sanctions," *Reuters*, 7 December 2014; Abbas Qaidaari, "More Planes, More Missiles, More Warships: Iran Increases Its Military Budget by a Third," *Al-Monitor*, 13 July 2015.
18 Bozorgmehr Sharafedin, "Iran to Expand Military Spending, Develop Missiles," *Reuters*, 9 January 2017; Bijan Khajehpour, "Will New 5-year Plan Solve Iran's Economic Problems?" *Al-Monitor*, 10 June 2016.
19 SIPRI's data on Iran.
20 SIPRI's data on Saudi Arabia is accessible through the World Bank's open data website.
21 "Iran Military Expenditure: 1988–2017," Trading Economics; Ben Moshinsky and Will Martin, "Saudi Arabia Spends £56 Billion a Year on Its Insane Military—Here's What That Kind of Money Buys," *Business Insider*, 19 February 2016; Craig Caffrey, "Saudi Arabia Reveals 6.6% Defence Spending Rise for 2017," *HIS Jane's 360*, 28 December 2016.
22 Benjamin D. Katz, "Saudi Arabia on Offensive with Attack-Minded Military Splurge," *Bloomberg*, 16 February 2017.
23 SIPRI's data on Saudi Arabia.
24 SIPRI's data on Israel is accessible through the World Bank's open data website.
25 S. Paul Kapur, *Jihad as Grand Strategy: Islamist Militancy, National Security, and the Pakistani State* (New York: Oxford University Press, 2017).
26 See: British Broadcasting Service, "Iran and the West: The Pariah State."
27 David Crist, *The Twilight War: The Secret History of America's Thirty-Year Conflict with Iran* (New York: Penguin Press, 2012); Scott Peterson, *Let the Swords Encircle Me* (New York: Simon & Shuster, 2010); Rouhollah K. Ramazani, *Revolutionary Iran: Challenge and Response in the Middle East* (Baltimore, MD: Johns Hopkins University Press, 1986).
28 BBC, "Iran and the West."
29 "Iranian Military to Fight for Lebanon," Tehran home service, June 7, 1982, *BBCSWB*, June 9, 1982.
30 "Iran: In Brief; Khameneh'i on training and help for Lebanon," Tehran home service, June 17, 1982, *BBC Summary of World Broadcasts*, June 19, 1982.
31 "Iran's Guards Rally to Beirut," *Newsweek*, June 28, 1982.
32 Kenneth Pollack, *The Persian Puzzle* (New York: Random House, 2004); William Beecher, "Identification of Iran's Factions May Provide Clues to Its Future," *Boston Globe*, November 16, 1986; Robin Wright, "Iranian Power Plays Reflected in Terrorist Moves?" *Christian Science Monitor*, November 5, 1986; Robin Wright, *In the Name of God: The Khomeini Decade* (New York: Simon & Schuster, 1989).
33 "Iran Revolution Guards' 'Liberation Movements Unit'," *BBC Summary of World Broadcasts*, February 6, 1981.
34 Nimatullah Ibrahimi, "The Failure of a Clerical Proto-State: Hazarajat, 1979–1984" (working paper series no. 2, Crisis States Research Center, London: Destin LSE, 2006);

"Confession by Captured Insurgent Trained in Iran," Kabul home service, March 17, 1985, *BBC Summary of World Broadcasts*, March 30, 1985.

35 Crist, *Twilight War*; Pollack, *The Persian Puzzle*; Wright, *In the Name of God*; Robin Wright, *Sacred Rage* (New York: Simon & Shuster, 1986).

36 Ronan Bergman, *The Secret War with Iran* (New York: Free Press, 2008); Trita Parsi, *Treacherous Alliance: The Secret Dealings of Israel, Iran, and the United States* (New Haven: Yale University Press, 2007); Nicholas Blanford, *The Warriors of God: Inside Hezbollah's Thirty-Year Struggle against Israel* (New York: Random House, 2011); Wright, *In the Name of God*, 108–109.

37 Owen Sirrs, *Iran's Quds Force: Proxy Wars, Terrorism, and the War on America* (Annapolis, MD: Naval Institute Press, 2022), 43.

38 H.E. Chehabi, *ed. Distant Relations: Iran and Lebanon in the last 500 years* (New York: St. Martin's Press, 2006), 180 –230; Haleh Vazeri, "Iran's Involvement in Lebanon: Polarization and Radicalization of Militant Islamic Movements," *Journal of South Asian and Middle Eastern Studies* 16, 2 (1992), 8.

39 Sayyed Ali Akbar Mohtashami, *Khaterat-e siyasi-e Sayyed Ali Akbar Mohtashami*, vol. 2 (Tehran: Khaneh-ye Andisheh-ye Javan, 2000), 108.

40 Yahya Rahim Safavi, in *Az jonub-e lobnan ta jonub-e iran: Khaterat-e sardar-e Sayyed Rahim Safavi*, ed. Majid Najafpour (Tehran: Markaz-e Asnad-e Enqelab-e Eslami, 2004), 95–107.

41 Ahmad Nizar Hamzeh, *In the Path of Hezbollah* (Syracuse: Syracuse University Press, 2004), 24–26.

42 Ibid., 25.

43 CIA, "Iran's Revolutionary Guard"; Christ, 126–127; Pollack, *Persian Puzzle*, 465–471.

44 Blanford, *Warriors of God*, 45 –46.

45 Crist, *The Twilight War*, 125.

46 Marc R. DeVore and Armin B. Stähli, "Explaining Hezbollah's Effectiveness: Internal and External Determinants of the Rise of Violent Non-State Actors," *Terrorism and Political Violence* 27, no. 2 (2015): 336–38, 342–45; Afshon Ostovar, "The Grand Strategy of Militant Clients: Iran's Way of War," *Security Studies* 28, no. 1 (2019): 8–11.

47 Ostovar, "The Grand Strategy of Militant Clients," 28.

48 Interview with Brother Mosleh, IRGC commander in Lebanon, *Payam-e Enqelab*, no. 138, June 8, 1985, 70.

49 Ibid., 70–71.

50 Crist, *Twilight War*, 128–129; Blanford, *Warriors of God*, 54; Hala Jaber, *Hezbollah: Born with a Vengeance* (New York: Columbia University Press, 1997), 112; Judith P. Harik, *Hezbollah: The Changing Face of Terrorism* (New York: I. B. Tauris, 2004), 170–172.

51 Bob Baer, *See No Evil* (New York: Crown, 2002).

52 Israeli Intelligence Officer quoted in: Bergman, *Secret War with Iran*, 66–67.

53 Crist, *The Twilight War*, 135–138, 150; Ronald Reagan, *ed. Radio Address to the Nation on Events in Lebanon and Grenada* (Washington, DC: The American Presidency Project, 1983); Ronald Reagan, *ed. Radio Address to the Nation on the Death of Federal Diplomatic and Military Personnel in Beirut, Lebanon* (Washington, DC: The American Presidency Project, 1983); Ronald Reagan, *ed. Radio Address to the Nation on the Situation in Lebanon* (Camp David, Maryland: The American Presidency Project, 1983).

54 Nicholas Cummins, "American Cold War Strategy and the Absence of 'Swift and Effective Retribution' for the 1983 Marine Barracks Bombing," *Diplomacy & Statecraft* 33, 2 (2022): 331–39.

55 DeVore and Stähli, "Explaining Hezbollah's Effectiveness," 346–48; Katzman, *Hizballah: A Radical Militia in Lebanon*, 11.

56 "Hariri Murder: UN Tribunal Issues Arrest Warrants," *BBC News*, June 30, 2011.

57 Pollack, *The Persian Puzzle*, 200–205; Jaber, *Hezbollah*, 112–115.

58 "Revolution Guards Offer Military Consultants," *Keyhan*, 8 March, 1990, in FBIS-NES-90-055, 21 March 1990.
59 Blanford, *Warriors of God*, 44–45.
60 Dexter Filkins, "The Shadow Commander," *The New Yorker*, September 23, 2013.
61 Meir Dagan cited in: Filkins, "The Shadow Commander."
62 Aymeric Janier, "La Force Al-Quds, Armeé de l'ombre du Régime," *Le Monde*, November 23, 2011; Meir Amit Center, *Using the Quds Force of the Revolutionary Guards as the Main Tool to Export the Revolution beyond the Borders of Iran* (Ramat Hasharon, Israel: Israel Intelligence Heritage and Commemoration Center, April 2, 2007).
63 Tom Coghlan, "Tehran Caught Arming Taleban with Weapons and Explosives," *Times* (London), March 18. 2010.
64 Meir Amit Center, *Using the Quds Force*; Blanford, *Warriors of God,* 341–342.
65 Amos Harel and Avi Issacharoff, *34 Days: Israel, Hezbollah, and the War in Lebanon* (New York: Palgrave Macmillan, 2008), 170.
66 Katzman, *Hizballah: A Radical Militia in Lebanon*, 9–10.
67 Kenneth Katzman, *Iran's Revolutionary Guard Named a Terrorist Organization*, CRS Report No. IN11093 (Washington, DC: Congressional Research Service, 2019), 1; Shannon W. Caudill, "Hizballah Rising: Iran's Proxy Warriors," *Joint Forces Quarterly* 49 (2008): 131.
68 Mark P. Sullivan and June S. Beittel, *Latin America: Terrorism Issues*, CRS Report No. RS21049 (Washington, DC: Congressional Research Service, 2015), 14.
69 Thanassis Cambanis, *A Privilege to Die: Inside Hezbollah's Legions and Their Endless War Against Israel* (New York: Free Press, 2010), 90–91; Blanford, *Warriors of God*, 199.
70 Debriefing of Detainee under Coalition Forces Control, Redacted Intelligence Report 003, derived from Combatting Terrorism Center at West Point, *Harmony Program*; Meir Amit Intelligence and Terrorism Information Center, *Ali Daqduq, Senior Lebanese Hezbollah Operative Handled by the Iranian Al-Quds Force to Support Shi'ite Militias in Iraq, Was Captured by the Coalition and Held by the Americans* (Ramat Hasharon, Israel: Israel Intelligence Heritage and Commemoration Center, January 4, 2012); Crist, *Twilight War*, 529–30; Steven K. O'Hern, *Intelligence Wars: Lessons from Baghdad* (New York: Prometheus Books, 2008), 106–7.
71 Ostovar, "The Grand Strategy of Militant Clients," 20–4.
72 Department of State, Country Reports on Terrorism: 2021 (Washington, DC: Department of State, 2021), 126, 146–7; Katzman, *Hizballah: A Radical Militia in Lebanon*, CRS Report No. 93-905 F (Washington, DC: Congressional Research Service, 1993), 9.
73 Blanford, *Warriors of God*, 44–48.
74 Carla E. Humud and Clayton Thomas, *Iran: Background and US Policy*, CRS Report No. R47321 (Washington, DC: Congressional Research Service, 2023), 14–5.
75 Adel Bin Ahmed al-Jubeir, "Can Iran Change?," *New York Times*, 19 January 2016; Afshon Ostovar, "Sectarianism and Iranian Foreign Policy," in *Beyond Sunni and Shia: The Roots of Sectarianism in a Changing Middle East*, ed. Frederick Wehrey (New York: Oxford University Press, 2018), 88–92.
76 John Irish and Andrea Shalal, "Saudi Arabia, Israel Present de facto United Front Against Iran," *Reuters*, 19 February 2017; Kapur, *Jihad as Grand Strategy*, 18–19, 27.
77 Kapur, *Jihad as Grand Strategy*, 18–19, 27.
78 Julian Borger and David Smith, "Trump Administration Imposes New Sanctions on Iran," *The Guardian*, 3 February 2017; Andrew Ward, "BP Opts Out of Iran Deals Ahead of Trump Hard Line on Tehran," *Financial Times*, January 2, 2017.
79 Kapur, *Jihad as Grand Strategy*, 10.
80 DeVore and Stähli, "Explaining Hezbollah's Effectiveness," 338.
81 Humud and Thomas, *Iran: Background and US Policy*, 9–10.
82 Ibid., CRS Report No. R47321 (Washington, DC: Congressional Research Service, 2023), 9–10, 19.
83 Jeffrey Feltman, "Hezbollah: Revolutionary Iran's most successful export," *Brookings*, January 17, 2019.

Bibliography

Al-Jubeir, Adel Bin Ahmed. "Can Iran Change?," *New York Times*, January 19, 2016, http://www.nytimes.com/2016/01/19/opinion/saudi-arabia-can-iran-change.html.

Arjomand, Said Amir. *The Turban for the Crown: The Islamic Revolution in Iran* (New York: Oxford University Press, 1988).

Arquilla, John, and David Ronfeldt. *Swarming and the Future of Conflict* (Santa Monica, CA: RAND Corporation, 2000).

Baer, Bob. *See No Evil* (New York: Crown, 2002).

Beecher, William. "Identification of Iran's Factions May Provide Clues to Its Future," *Boston Globe*, November 16, 1986.

Bergman, Ronan. *The Secret War With Iran* (New York: Free Press, 2008).

Bijan Khajehpour. "Will New 5-year Plan Solve Iran's Economic Problems?," Al-Monitor, June 10, 2016, https://www.al-monitor.com/originals/2016/06/iran-sixth-five-year-development-plan-rouhani-parliament.html.

Blanford, Nicholas. *Warriors of God: Inside Hezbollah's Thirty-Year Struggle Against Israel* (New York: Random House, 2011).

Borger, Julian, and David Smith. "Trump Administration Imposes New Sanctions on Iran," *The Guardian*, February 3, 2017, https://www.theguardian.com/us-news/2017/feb/03/trump-administration-iran-sanctions.

Bozorgmehr Sharafedin. "Iran to Expand Military Spending, Develop Missiles," *Reuters*, January 9, 2017, http://www.reuters.com/article/us-iran-military-plan-idUSKBN14T15L.

British Broadcasting Service. "Iran and the West: The Pariah State," July 4, 2009. https://www.bbc.co.uk/programmes/b00ht3p7.

Caffrey, Craig. "Saudi Arabia Reveals 6.6% Defence Spending Rise for 2017," *HIS Jane's 360*, December 28, 2016, https://www.janes.com/osint-insights/defence-news/saudi-arabia-cuts-defence-spending.

Cambanis, Thanassis. *A Privilege to Die: Inside Hezbollah's Legions and Their Endless War Against Israel* (New York: Free Press, 2010).

Caudill, Shannon W. "Hizballah Rising: Iran's Proxy Warriors," *Joint Forces Quarterly* 49 (2008): 128–134.

Chehabi, H.E., ed. *Distant Relations: Iran and Lebanon in the Last 500 Years* (New York: St. Martin's Press, 2006).

Coghlan, Tom. "Tehran Caught Arming Taleban with Weapons and Explosives," *Times*, March 18, 2010.

"Confession by Captured Insurgent Trained in Iran," Kabul home service, March 17, 1985, *BBC Summary of World Broadcasts*, March 30, 1985.

Crist, David. The Twilight War: The Secret History of America's Thirty-Year Conflict With Iran (New York: Penguin Books, 2013).

Cummins, Nicholas. "American Cold War Strategy and the Absence of 'Swift and Effective Retribution' for the 1983 Marine Barracks Bombing," *Diplomacy & Statecraft* 33, no. 2 (2022): 330–354.

Debriefing of Detainee under Coalition Forces Control. Redacted Intelligence Report 003, derived from Combatting Terrorism Center at West Point, *Harmony Program* (2007), https://www.ctc.usma.edu/v2/wp-content-uploads/2013/10/Redacted-Intelligence-Report-003-Summary.pfd.

Department of State. Country Reports on Terrorism: 2021. Washington, DC: Department of State, 2021, https://www.state.gov/reports/country-reports-on-terrorism-2021/.

DeVore, Marc R., and Armin B. Stähli. "Explaining Hezbollah's Effectiveness: Internal and External Determinants of the Rise of Violent Non-State Actors," *Terrorism and Political Violence* 27, no. 2 (2015): 331–57.

Feltman, Jeffrey. "Hezbollah: Revolutionary Iran's most successful export," Brookings, January 17, 2019, https://www.brookings.edu/articles/hezbollah-revolutionary-irans-most-successful-export/.

Filkins, Dexter. "The Shadow Commander," *The New Yorker*, September 23, 2013, https://www.newyorker.com/magazine/2013/09/30/the-shadow-commander.
Hamzeh, Ahmad Nizar. *In the Path of Hezbollah* (Syracuse: Syracuse University Press, 2004).
Harel, Amos, and Avi Issacharoff. *34 Days: Israel, Hezbollah, and the War in Lebanon* (New York: Palgrave Macmillan, 2008).
Harik, Judith P. *Hezbollah: The Changing Face of Terrorism* (New York: I. B. Tauris, 2004).
"Hariri murder: UN tribunal issues arrest warrants," BBC News, June 30, 2011, https://www.bbc.com/news/world-middle-east-13972350.
Humud, Carla E, and Clayton Thomas. *Iran: Background and US Policy*. CRS Report No. R47321. Washington, DC: Congressional Research Service, 2023, https://congressional-proquest-com.libproxy.nps.edu/congressional/docview/t21.d22.crs-2022-crs-228627?accountid=12702.
Ibrahimi, Nimatullah. *The Failure of a Clerical Proto-State: Hazarajat, 1979-1984*" (working paper series no. 2, London: Destin LSE: Crisis States Research Center, 2006).
Interview with Brother Mosleh, IRGC commander in Lebanon. *Payam-e Enqelab*, no. 138, June 8, 1985.
"Iran's Guards Rally to Beirut," *Newsweek*, June 28, 1982.
"Iranian Revolutionary Guards," Military Communique," Tehran home service, October 30, 1980, *BBC Summary of World Broadcasts*, November 1, 1980.
"Iranian Military to Fight for Lebanon," Tehran home service, June 7, 1982, *BBCSWB*, June 9, 1982.
"Iran: In Brief; Khameneh'i on training and help for Lebanon," Tehran home service, June 17, 1982, *BBC Summary of World Broadcasts*, June 19, 1982.
"Iran Revolution Guards' 'Liberation Movements Unit'," *BBC Summary of World Broadcasts*, February 6, 1981.
"Iran Military Expenditure: 1988–2017," Trading Economics, http://www.tradingeconomics.com/iran/military-expenditure.
Irish, John, and Andrea Shalal. "Saudi Arabia, Israel Present de facto United Front Against Iran," *Reuters*, 19 February 2017, http://www.reuters.com/article/us-mideast-crisis-iran-idUSKBN15Y09R.
Jaber, and Hala. *Hezbollah: Born With a Vengeance* (New York: Columbia University Press, 1997).
Janier, Aymeric. "La Force Al-Quds, armeé de l'ombre du régime," *Le Monde*, November 23, 2011.
Kapur, S. Paul. *Jihad as Grand Strategy: Islamist Militancy, National Security, and the Pakistani State* (New York: Oxford University Press, 2017).
Katz, Benjamin. "Saudi Arabia on Offensive With Attack-Minded Military Splurge," *Bloomberg*, February 16, 2017, https://www.bloomberg.com/news/articles/2017-02-17/saudi-arabia-on-offensive-with-attackmindedmilitary-splurge.
Katzman, Kenneth. *The Warriors of Islam: Iran's Revolutionary Guards* (New York: Routledge, 1993).
Katzman, Kenneth. *Hizballah: A Radical Militia in Lebanon*. CRS Report No. 93-905 F. Washington, DC: Congressional Research Service, 1993).
Katzman, Kenneth. *Iran's Revolutionary Guard Named a Terrorist Organization*, CRS Report No. IN11093 (Washington, DC: Congressional Research Service, 2019), 1, https://crsreports.congress.gov/product/pdf/IN/IN11093.
Maher, Andrew. "The Middle Eastern Advisory Competition: The Islamic Revolutionary Guards Corps Quds Force." Australian Army Journal XIV, 1 (Autumn 2018): 57–82.
Mahfoozi, Alireza. in an interview recorded by Zia Shedghi, April 7, 1984. Paris, France. Iranian Oral History Collection, Harvard University.
Meir Amit Center. *Using the Quds Force of the Revolutionary Guards as the Main Tool to Export the Revolution Beyond the Borders of Iran* (Ramat Hasharon, Israel: Israel Intelligence Heritage and Commemoration Center, 2007).
Meir Amit Intelligence and Terrorism Information Center. *Ali Daqduq, Senior Lebanese Hezbollah Operative Handled by the Iranian Al-Quds Force to Support Shi'ite Militias in*

Iraq, Was Captured by the Coalition and Held by the Americans (Ramat Hasharon, Israel: Israel Intelligence Heritage and Commemoration Center, January 4, 2012).

Mohtashami, Sayyed Ali Akbar. *Khaterat-e Siyasi-e Sayyed Ali Akbar Mohtashami 2* (Tehran: Khaneh-ye Andisheh-ye Javan, 2000).

Moshinsky, Ben, and Will Martin. "Saudi Arabia Spends £56 Billion a Year on Its Insane Military—Here's What That Kind of Money Buys," *Business Insider*, February 19, 2016. https://www.businessinsider.in/saudi-arabia-spends-16356-billion-a-year-on-its-insane-military-heres-what-that-kind-of-money-buys/articleshow/51063160.cms.

O'Balance, Edgar. *The Gulf War* (London: Brassey's, 1988).

O'Hern, and K Steven. *Intelligence Wars: Lessons from Baghdad* (New York: Prometheus Books, 2008).

Ostovar, Afshon. *Vanguard of the Imam, Religion, Politics, and Iran's Revolutionary Guards* (New York: Oxford University Press, 2016).

Ostovar, Afshon. "Sectarianism and Iranian Foreign Policy," in *Beyond Sunni and Shia: The Roots of Sectarianism in a Changing Middle East*, ed. Frederick Wehrey, 88–92. (New York: Oxford University Press, 2018).

Ostovar, Afshon. "The Grand Strategy of Militant Clients: Iran's Way of War," *Security Studies* 28, no. 1 (2019): 159–88.

Parsi, Trita. *Treacherous Alliance: The Secret Dealings of Israel, Iran, and the United States* (New Haven: Yale University Press, 2007).

Payam-e Enqelab. no 60, June 12, 1982.

Payam-e Enqelab. no. 160.

Payam-e Enqelab. no. 161.

Peterson, Scott. *Let the Swords Encircle Me* (New York: Simon & Shuster, 2010).

Pollack, Kenneth. *The Persian Puzzle* (New York: Random House, 2004).

Ramazani, Rouhollah K. *Revolutionary Iran: Challenge and Response in the Middle East* (Baltimore, MD: Johns Hopkins University Press, 1986).

Reagan, Ronald, ed. Radio Address to the Nation on Events in Lebanon and Grenada. Washington, DC: The American Presidency Project, 1983. https://www.presidency.ucsb.edu/documents/address-the-nation-events-lebanon-and-grenada.

Reagan, Ronald, ed. Radio Address to the Nation on the Death of Federal Diplomatic and Military Personnel in Beirut, Lebanon. Washington, DC: The American Presidency Project, 1983, https://www.presidency.ucsb.edu/documents/radio-address-the-nation-the-death-federal-diplomatic-and-military-personnel-beirut.

Reagan, Ronald, ed. Radio Address to the Nation on the Situation in Lebanon. Camp David, Maryland: The American Presidency Project, 1983, https://www.presidency.ucsb.edu/documents/radio-address-the-nation-the-situation-lebanon.

"Revolution Guards Offer Military Consultants," *Keyhan*, March 8, 1990, in FBIS-NES-90-055, March 21, 1990.

Safavi, Yahya Rahim. *Az Jonub-e Lobnan Ta Jonub-e Iran: Khaterat-e Sardar-e Sayyed Rahim Safavi*, ed. Majid Najafpour (Tehran: Markaz-e Asnad-e Enqelab-e Eslami, 2004).

Shaery-Eisenlohr, Roschanack. *Shi'ite Lebanon: Transnational Religion and the Making of National Identities* (New York: Columbia University Press, 2008), 94–99.

SIPRI's data on Iran is accessible through the World Bank's Open Data website. http://data.worldbank.org/indicator/MS.MIL.XPND.GD.ZS?locations&hx003D;IR.

SIPRI's data on Saudi Arabia is accessible through the World Bank's Open Data website. http://data.worldbank.org/indicator/MS.MIL.XPND.GD.ZS?locations&hx003D;SA.

Sirrs, Owen. *Iran's Quds Force: Proxy Wars, Terrorism, and the War on America* (Annapolis, MD: Naval Institute Press, 2022).

Sullivan, Mark P., and June S. Beittel. Latin America: Terrorism Issues. CRS Report No. RS21049. Washington, DC: Congressional Research Service, 2015, https://congressional.proquest.com/congressional/result/congressional/congdocumentview?accountid=12702&groupid=100340&parmId=187DDA412F3&rsId=187DD9AB36C#0.

"US Embassy Held by Iranian Students in Bid to Get Shah," The Globe and Mail. November 5, 1979.

Ward, Andrew. "BP Opts Out of Iran Deals Ahead of Trump Hard Line on Tehran," *Financial Times*, 2 January 2017, https://www.ft.com/content/b73c0e9e-ce7b-11e6-b8ce-b9c03770f8b1.

Wright, Robin. "Iranian Power Plays Reflected in Terrorist Moves?" *Christian Science Monitor*, November 5, 1986.

Wright, Robin. *Sacred Rage* (New York: Simon & Shuster, 1986).

Wright, Robin. *In the Name of God: The Khomeini Decade* (New York: Simon & Schuster, 1989).

Vazeri, Haleh. "Iran's Involvement in Lebanon: Polarization and Radicalization of Militant Islamic Movements," *Journal of South Asian and Middle Eastern Studies* 16, 2 (1992).

Zabih, Sepehr. *The Iranian Military in Revolution and War* (New York: Routledge, 1988).

8 China's Shaping Operations in the South China Sea

Introduction

Just as the Islamic Revolutionary Guard Corps (IRGC) was able to operate below the detection threshold and continue a long-term operation in Lebanon by exploiting social and cultural fault lines, the People's Republic of China (PRC) has been able to wage a protracted, irregular warfare campaign in the South China Sea (SCS) by exploiting the ambiguous character of its maritime militia, a paramilitary capability.

The campaign aims to defend and advance several of China's national interests, including restoring territorial integrity, as interpreted by the Chinese government. The militia forms the tip of this campaign's spear by performing special warfare. This allows China to achieve its goals while making attribution and, by extension, delivering effective and timely responses problematic. The continued and discrete deployment of the militia illustrates its strategic utility, particularly in peacetime, as its ambiguous character provides the Chinese government with a broader set of strategic choices to achieve political objectives.

Background

China has six national interests that interconnect, complement, or even conflict with one another. These six national interests are state sovereignty, national security, territorial integrity, national reunification with Taiwan, protecting the Chinese Communist Party (CCP) and overall social stability, and basic safeguards for ensuring sustainable economic and social development.[1] To advance its national interests of state sovereignty, national security, and territorial integrity, China makes several territorial claims in the SCS.[2] As a result, China has ongoing disputes over islands and maritime territory boundaries in the area with six other countries, including Brunei, Indonesia, Malaysia, the Philippines, Taiwan, and Vietnam.[3]

The United Nations Convention on the Law of the Sea (UNCLOS), which China signed in 1982 and ratified in 1996, lies at the core of the conflicts. UNCLOS regulates the use of the oceans and their resources.[4] The current disputes in the SCS, such as territorial claims, fishing grounds, freedom of navigation, or maritime rights of (underwater) features, are all related to the UNCLOS agreements.[5]

DOI: 10.4324/9781003365112-11

In 2016, the Permanent Court of Arbitration (PCA) stated that China's claims were not compliant with UNCLOS.[6] The disputes over the SCS thus form the core of China's challenge to the international rule-based order.

Over the last decade, tensions over several disputed areas in the SCS have increased. Although these disputes are local, the areas in which they occur are of strategic importance to international security and global trade, as an estimated $3.37 trillion worth, or twenty-one percent of all global trade, transited through the area in 2016.[7] Thus, the SCS is a vital maritime shipping route for many of the world's leading economies. In 2016, for example, 64 percent of China's maritime trade and 42 percent of Japan's maritime trade transited through the SCS in the same year. In 2023, that level of international trade continued to cross the same area. The United States' (US) dependence on the waterway is less significant, as just over 14 percent of American maritime trade passed through the waterway in 2016.[8]

Of all the countries involved in the disputes, China claims the most significant portion of the SCS, including features in the other countries' exclusive economic zones (EEZs). According to People's Liberation Army Navy (PLAN) Admiral (Ret.) Liu Huaqing, this area, known as the First Island Chain, encompasses "the wide sea areas west of the Japanese archipelago, the Ryukyu Islands, and the Philippine islands."[9] Chinese vessels must pass through a series of straits and channels that lay between this group of islands to make their way from Chinese ports into the Pacific Ocean. China considers twenty-two of these straits and channels to be vital to its national security and economic development.[10] Figure 8.1 presents a map of the South China Sea with the First Island Chain.

China considers the First Island Chain a potential threat to its national security. First, the straits and channels that China considers to be vital to its national security and economic development are under the control of other states. As a result, China claims that its maritime trade could become subject to a naval blockade. Second, the island chain forms a clear demarcation line around an area rich in natural resources. For example, the area contains an estimated eleven billion barrels of oil, one hundred and ninety trillion cubic feet of natural gas, and more than sixteen million tons of annual fishing catch.[11] The area's economic value thus provides a significant incentive to claim economic maritime rights. From a military perspective, long-range, precision-strike capabilities emplaced along the island chain could reach China's large coastal cities. China's major political and economic hubs are situated along its eastern coast, while most of the islands in the western Pacific are under the control of Japan, the US, or their partners. Therefore, China believes it lacks the strategic depth to effectively defend these vital coastal areas in case of armed conflict.[12] To counter this unfavorable balance of power and exploit economic opportunities, gaining control of the SCS is a vital Chinese national interest. To this end, China seized several islands in the area from 1975 to 2019, built artificial ones, and constructed military installations such as bases, airfields, and harbors to enhance its forward military presence. By doing so, China intends to decrease its sense of insecurity, establish a buffer zone, and improve the People's Liberation Army's (PLA) reach in case of aggression.[13]

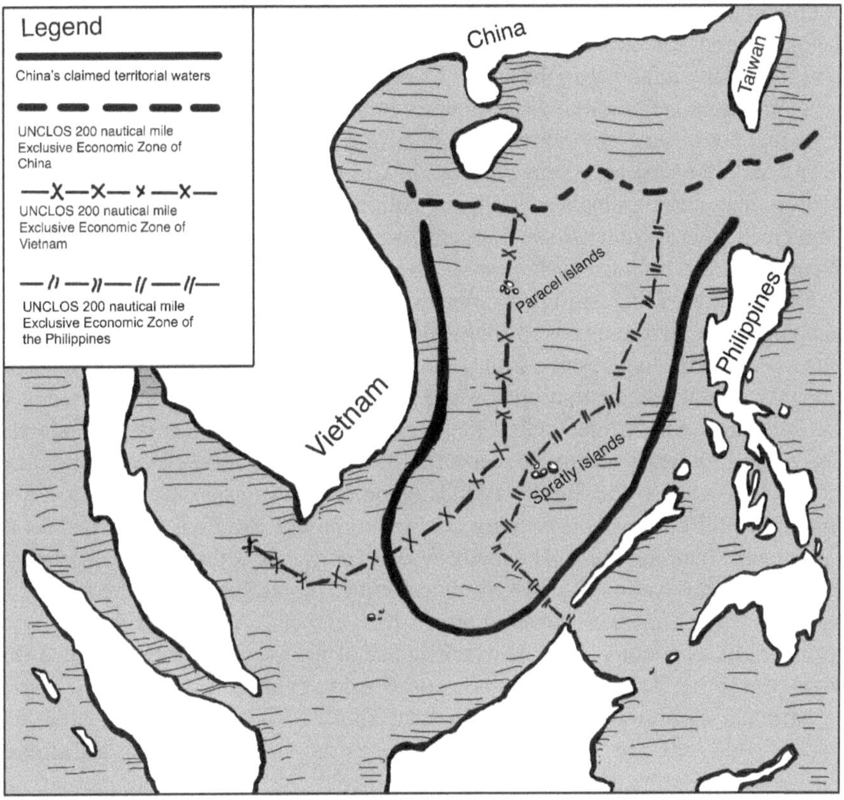

Figure 8.1 The South China Sea (SCS) with the First Island Chain.

The US has growing concerns about China's activities in the SCS.[14] The military bases built over the last decade increase the Chinese military footprint in the region, which could offset US military capabilities. Indeed, China has reinforced its Anti-Access-Area-Denial (A2/AD) capabilities, which are a significant challenge for US military operations in the area.[15] For example, this changing balance of military force threatens the security guarantees the US provides to Taiwan, the Philippines, and South Korea, as well as the US's ability to intervene militarily in case of a crisis over Taiwan. Furthermore, off-setting US military capabilities might persuade other countries in the region to rethink their own national security policy and defense strategy, including military partnerships.[16] Thus, the balance of power in the SCS is strongly related to contemporary hegemonic rivalry known as strategic competition.

As early as 1974, China employs a protracted irregular warfare campaign to gain control of the SCS. The campaign includes seizing territory and reclaiming land, among other things, to expand its footprint and enhance its forward military presence. Coercive actions are a calculated response to other actors in which costs – mainly relational – serve as a balancing factor that characterizes this campaign.

On the one hand, China lays more aggressive maritime claims in the area. On the other hand, it's aware of the various regional relationships and suspicions due to its increasing influence in the area.[17] China is, therefore, balancing its actions and policy to minimize possible disadvantages in its regional or international relationships that could seriously harm its long-term goals. This balancing approach is a clear example of the interconnectedness of China's national interests, as claiming territory may serve China's sovereignty, national security, and territorial integrity. Conversely, increased tensions, or even armed conflicts with other claimants, may threaten China's political system established by its constitution.

Five periods of fluctuating tensions and relationships stand out in this extended campaign. The first and most prolonged period stretched from 1974 to 2002 and witnessed the most aggressive Chinese actions in the SCS. In 1974, Chinese forces violently seized the western part of the Paracel Islands from South Vietnam. In the following years, China constructed a military infrastructure at Woody Island.[18] The second aggressive move occurred in 1995. The Chinese Navy took control of Mischief Reef in the Spratly Islands, another island group in the SCS. China seized the reef after a short military confrontation with the Philippines and set up a permanent base structure. The incident led to concerns within the Association of Southeast Asian Nations (ASEAN) about further destabilization of the region. In November 2002, tensions lowered after China and ASEAN signed the Treaty of Amity and Cooperation (TAC), which included the commitment to resolve disputes peacefully, but the Chinese base continued to exist.[19]

After the signing of the TAC, a relatively short, stable period (2002–2006) followed.[20] But, between 2006 and 2008, tensions increased again after a Chinese–Vietnamese dispute over offshore development. During this third period, the relationships in the region further deteriorated. In May 2009, the UN's Commission on the Limits of the Continental Shelf (CLCS) incentivized all claimants to substantiate their claims.[21] Although not explicitly, China introduced its historic – and infamous – claim illustrated by the nine-dash line (9DL), which encompasses almost ninety percent of the SCS.[22] Different incidents followed, the most prominent being the Chinese harassment of the USNS *Impeccable*. In 2010, the US rejected the 9DL and strengthened regional partnerships. The increased US regional involvement led China to moderate its approach to limit the "internationalization of the disputes."[23]

The US involvement also led to renewed tensions in the region. In April 2012, a Philippine naval vessel arrested members of the militia at Scarborough Shoal. This resulted in a two-month confrontation between Philippine and Chinese vessels. Eventually, both nations removed their ships, but China immediately took control of the shoal. China continued its assertive approach by sending the militia to the disputed areas. China also strengthened its military garrison in Sansha City and adjusted its administrative status, thereby increasing its jurisdiction over the area.[24] After the standoff, China made several efforts to emphasize cooperation instead of disputes. China initiated talks with ASEAN, established a hotline with the Vietnamese Navy, and China's President Xi Jinping visited all claimants in the SCS, except the Philippines.[25]

The fourth period of tensions started in 2014. China constructed landmasses on the Spratly Islands, including a harbor and airstrip at Fiery Cross Reef. This marked the period of island building in the SCS, which the US judged to be one of the most provocative Chinese actions.[26] A second significant Chinese provocation was the deployment of the HYSY-981 oil drilling rig near the Paracel Islands in Vietnam's EEZ. This led to a strong ASEAN reaction and tightened US-Vietnam relationships.[27] In response, China moderated its approach again by offering loans worth billions of US dollars for infrastructural developments in the region and reassuring agreements between Xi and US President Obama in November 2014.[28]

The fifth and last period ran from 2015 to the end of 2022. Although China continues land reclamation and other provocations, this period has not seen a significant behavioral change sparkling further escalation. In 2018, China successfully landed an H-6 long-range bomber on Woody Island. In 2020, China emplaced anti-ship cruise missiles and surface-to-air missile systems on the Spratly Islands.[29] Despite the militarization of the islands and US concerns, the situation in the SCS seemed to have stabilized.[30] China also made some progress on bilateral engagements by establishing a hotline with Japan and warming ties with Philippine President Duterte.[31]

On the day of this writing, China's coercive actions are growing again, not only in number but also in intensity. A substantial number of incidents in 2023 occurred near the Second Thomas Shoal, where the China Coast Guard (CCG) and militia vessels prevented the Philippine Navy and Coast Guard from resupplying the BRP *Sierra Madre*. This old US Navy warship serves as an outpost at the shoal. Although China is the aggressor, the Philippine government, now led by President Ferdinand Marcos Jr., has (in China's view) taken several escalatory steps, including the emplacement of navigation buoys, increasing the Philippine Coast Guard's presence, and exposing China's activities in the media (i.e., transparency initiative), triggering a Chinese reaction. The US has also become more active in 2023 by organizing senior leader engagements in the region, employing surveillance aircraft supporting the Philippine military, and issuing strong statements to condemn the Chinese actions while reiterating its support to the Philippines.

A short examination of the disputes in the SCS shows that China has been consistently pursuing its maritime claims since 1974. As it oscillates between intensifying provocations and repairing the subsequently damaged relationships, China has nevertheless improved its strategic position in the SCS. In 2024, China has a strong presence in the area, including several islands on which the PLA has permanent infrastructure and forces. Furthermore, China has not diminished its territorial claims. So, supported by a well-defined policy and remarkably following the 9DL, China has improved its position in the SCS and occupied the territory of other regional actors, including the Philippines. However, these achievements have come at a price for China's international image, including the rise of alliances to counter China's ambitions. This shows that China's balanced approach between security and relationships has not achieved an optimal equilibrium: the physical aims in the SCS came at a cost.

Despite these high costs, China upholds its historical claims and is willing to take risks to defend its perceived rights. To mitigate risks, China's strategy includes using various military and non-military actors to stay below the threshold of armed escalation. What stands out is the use of the maritime militia. This militia is an example of one of China's discrete capabilities to expand its control of the SCS, often in conjunction with the CCG and regular Navy. In doing so, the militia allows China to apply coercion and presents an emerging twenty-first century security threat in an era of rising strategic competition over the SCS.

China's maritime militia

China is home to the world's largest fishing fleet, consisting of between 12,000 and 17,000 vessels and an even more significant number of people who work either on the vessels themselves or in the related onshore industries.[32] A substantial portion of these vessels and their crews are members of the maritime militia. The people forming this militia are essentially military personnel and trained as such. The militia is prepared to defend and advance China's national interests, including expanding its maritime territorial claims in the SCS. According to Kennedy and Erickson, the militia may also be deployed to support the regular Navy during wartime.[33]

From an organizational perspective, the maritime militia is part of the Militia of China, which is part of the Chinese military. A 2013 Chinese Defense White Paper describes the Militia of China as "an assistant and backup force of the PLA [People's Liberation Army]."[34] Because the Militia of China's members retain their day-to-day civilian jobs and only participate in training, exercises, or operations on demand, the militia should be considered a reserve force operating in parallel with or supporting the army. To this end, the Militia of China has three distinct tasks: assisting the Chinese military in defending China's homeland, helping local authorities maintain order and stability, and providing disaster relief.[35] Similar to traditional Chinese military units, the militia is characterized by a military organizational structure. Due to their dual tasks, civil-military leadership exercises so-called dual-track command-and-control over the Militia of China.[36]

The maritime militia, together with the regular Navy, plays a substantial role in China's joint military, law enforcement, and civilian defense efforts at sea.[37] These joint efforts allow the effective coordination of the militia with traditional military components. To further enhance interoperability, the most advanced units of the maritime militia receive tactical military training from the uniformed military.[38] Furthermore, the militia participates in joint exercises with the regular Navy and Coast Guard.[39]

The militia, thus, enjoys a blurred status that allows them to operate discretely, deceive potential opponents, and significantly lower the risk of escalation. This ambiguous status particularly applies to some of its most advanced units that collect intelligence and conduct reconnaissance.[40] These advanced units make up only a small portion of the militia but are the ones that are involved in important, ambiguous activities far from mainland China. The militia's most professionalized units are equipped with various high-level intelligence, surveillance, and reconnaissance (ISR) capabilities and operate from China's southernmost province, Hainan, and include the Danzhou, Tanmen, Sanya, and Sansha militias. All four militias played a significant role in several key events that took place over the last five decades in the SCS that were described above.

The maritime militia's activities

The operational preparation of the environment (OPE) is a useful lens through which to analyze and describe the maritime militia's special warfare activities. US military doctrine describes OPE as "non-intelligence activities [i.e., Title 10 military authorities] conducted to plan and prepare for potential follow-on military operations."[41] The goal of OPE is to allow military units to become "familiar with the area in which [these units] might have to work."[42] In other words, OPE aims to enhance military forces' situational awareness and understanding of the three dimensions (i.e., physical, virtual, and cognitive) of the Operational Environment (OE) in support of future planning and operations.

US military doctrine lists three critical activities as part of OPE. The first activity is orientation activities that focus on conducting area assessments and site surveys, as well as identifying the most suitable routes to targets and the potential obstacles that forces may encounter during infiltration and exfiltration. The second activity is target development, which aims to collect real-time information on a specific target to support operational planning and future operations. Key characteristics of target development are exploiting local knowledge of a geographical area or specific expertise and using ISR capabilities to find and fix targets. The final activity is the preliminary engagement of a target. Preliminary engagement aims to find, fix, track, monitor, and influence a target by employing persistent surveillance or conducting small-scale offensive actions before follow-on operations by a main force.[43]

Next to this functional distinction, military concepts often separate OPE in time. When conducted in peacetime, the US military describes OPE as pre-crisis activities. During wartime, OPE is known as advanced force operations (AFO), which specially selected personnel conduct immediately before the employment of the main forces.[44] The militia conducts all three activities considered part of OPE in conjunction with the CCG and regular Navy to support China's irregular warfare strategy in the SCS.

Orientation activities

Most of the militia's operations are orientation activities. During the first six months of 2023, the Philippine Coast Guard (PCG) observed a significant increase in the militia operating in the SCS. The number of Chinese ships in that period totaled seven hundred compared to five hundred in the same timeframe in 2022. Approximately 85 percent of these Chinese vessels are part of the militia. Most of these vessels swarmed in strategic areas such as the Spratly Islands (see Figure 8.1).

To continue their presence and maintain the desired level of awareness, the militia rotates its vessels in conjunction with CCG vessels and Navy ships. The Coast Guard vessels protecting the entrances of specific locations, such as Scarborough Shoal, typically rotate on an eight-week basis. In contrast, the ones that patrol a more general area rotate every four weeks. The so-called fishing vessels of the militia inside these specific locations remain present for extended periods. Notably, the PCG noticed a significant change in the total period the militia's vessels were present in the SCS. In 2022, deployments were typically one month long; in 2023, that became five months. In addition, the militia applies swarming tactics of their vessels that center on employing many small and dispersed flotillas that form a networked force.[45] Using these tactics over extensive periods of time allows the militia to cover a large area with a limited number of vessels, enhancing its situational awareness and understanding of crucial parts of the SCS, while decreasing the chances of detection.

Employing swarming tactics requires a different way of thinking about organizational structure and the definition and application of the elements of combat power. Because of the militia's ability to maintain encrypted communications among each other, and with CCG vessels and Navy ships, as well as with forward bases in the SCS and mainland China, they function as a force multiplier that provides the Chinese military with an information advantage achieved promptly and at low cost, particularly when compared to the employment of regular Navy vessels. Regarding continuity, the militia's protracted deployments combined with its swarming tactics indicate that other Chinese ships are likely responsible for providing logistics support from mainland China, as fishing vessels generally can only stay at sea for several weeks before they need resupplies. Assuming the existence of these supply chains, this would be a strong indicator that more robust command-and-control arrangements between the Chinese Navy, Coast Guard, and militia are in place to sustain China's irregular warfare campaign in the SCS.

Target development

The militia conducts target development in support of Chinese coercive actions and the reclamation of territory. The "fishermen" deeply understand the SCS due to their long-term experience on the sea. They switch off the transmitting component of their Automatic Identification System (AIS) transponder, thus only receiving data on specific targets to support the Chinese military's operational planning while concealing their own position. For example, on August 2, 2023, several militia vessels prepositioned themselves discreetly near the Second Thomas Shoal. By doing so, they could conduct surveillance and perform an early warning function. Three days later, on August 5, the Filipinos sailed to the Shoal to complete their rotation and resupply mission of the *Sierra Madre*, grounded at the Shoal. The Philippine vessels included two Coast Guard vessels, two Navy ships, and two civilian resupply boats. Almost immediately after the Philippine vessels made their way to the shoal, twenty Chinese ships, twelve of which were militia vessels, six Coast Guard, and four regular Navy, entered the scene. The Chinese also deployed a helicopter.

The timely manner in which this force package responded illustrates the significance of discrete militia surveillance, as it allowed the Chinese to prepare themselves from over the horizon and remain on standby until alerted by the militia.

The Chinese vessels coordinated their actions smoothly, suggesting a robust command-and-control structure. The fact that an on-scene commander directed the operation from over the horizon, either at sea or from Mischief Reef, supports this assumption. For example, the Chinese vessels prepositioned themselves strategically in the hours before the arrival of the Filipinos. Their early presence suggests a Chinese information advantage, likely achieved through superior intelligence collection. During the Philippine mission, the Chinese vessels also conducted various maneuvers sequentially or simultaneously, showing tactical coordination. The Chinese combined these maneuvers with distracting messages to further deceive the Philippine crews. At first, these broadcasts were transmitted in Mandarin, followed by English. The Philippine military assessed that China uses these information operations as an integral part of its TTPs to achieve its goals, inform its domestic audience, and prevent potential internal division within the Chinese population about its coercive actions in the area.

Ultimately, the Chinese vessels blocked the Philippine resupply vessels by taking position on the starboard and port side, as well as the rear, at a distance as close as forty yards. By doing so, the Chinese forced the civilian resupply vessel in a direction other than toward the Shoal, thereby disrupting the Philippine activities, which is vital for sustaining the presence of Philippine military forces on board the *Sierra Madre*.

Preliminary engagement

The militia preliminarily engages targets. For example, on April 8, 2012, a Philippine Long Range Patrol Aircraft (LRPA) conducted a routine reconnaissance flight over the archipelagic country's EEZ. While approaching Scarborough Shoal, a triangle-shaped chain of reefs and rocks located approximately two-hundred-twenty kilometers northwest of the Philippines's Zambales province, the aircraft's crew spotted eight militia vessels belonging to the Tanmen militia anchored inside the shoal. The Philippine authorities considered the presence of these vessels in their EEZ an act of illegal fishing and sent its largest naval ship, the BRP *Gregorio del Pilar* (PF-15), to the shoal.

Two days later, the *Gregorio* arrived at the shoal and launched a boarding team to search one of the Chinese vessels. During the inspection, the team found numerous corals, giant clams, and live sharks that were all collected illegally. It was not the first time that Chinese fishers violated the International Convention on International Trade in Endangered Species of Wild Fauna and Flora (CITES), as the Philippine authorities had arrested several of them in the past for violating Philippine sovereignty and illegally fishing.

During the boarding, however, the captain of fishing vessel *Qionghai* 02096 (which was anchored inside the shoal) sent several messages to the People's Armed Police Border Defense Force Control Station situated in Hainan province.

Two civilian China Marine Surveillance vessels, the *Zhonggou Haijian* 75 and *Zhonggou Haijian* 84, that were near the shoal, responded to the distress call and set sail to the shoal. Just when the boarding team of the *Gregorio* wanted to arrest the "Chinese fishers," the two surveillance vessels arrived at the shoal and took position between the Chinese fishing vessels and the *Gregorio*. The captain of one of the Chinese vessels then instructed the captain of the *Gregorio* to leave the shoal immediately because he had entered Chinese territorial waters illegally.

The Philippine authorities initially responded to the incident by seeking a diplomatic solution. The administration further emphasized that they had not requested US assistance then. To deescalate the situation, the Philippine Navy withdrew the *Gregorio* from the scene on April 12 and replaced it with the BRP *Pampanga*, a search and rescue Coast Guard vessel. Rather than responding in a similar de-escalatory manner, the Chinese authorities deployed the *Yuzheng*-310, its latest fishery patrol vessel, to complement the two surveillance vessels already at the shoal. Additionally, the Chinese Ministry of Foreign Affairs issued a press release in which they stated that "the attempt by the Philippines to carry out so-called law enforcement activities in the waters of Huangyan Island has infringed upon China's sovereignty and runs counter to the consensus reached by both sides on maintaining the peace and stability in the South China Sea (SCS) and not to complicate and escalate the situation."[46]

The following day, Philippine foreign secretary Alberto del Rosario met with the Chinese ambassador in Manilla to start negotiations over the ongoing dispute. Shortly after the talks had begun, two Chinese cutters escorted the militia's vessels out of the shoal, leaving each country one vessel. When evening came, the negotiations stalled because the Philippine authorities had strongly objected to allowing the militia's vessels to leave the shoal with their catch untouched. The Chinese, on their part, insisted the Philippines would withdraw its last vessel from the lagoon. According to del Rosario, this Chinese request was unacceptable, and he announced a stalemate. Shortly after that, a second Chinese vessel arrived at the shoal.

After five days without diplomatic breakthroughs, the Philippine authorities changed their strategy by officially announcing that they sought international assistance to solve the dispute. In addition to this public call, the Philippine government formally requested that ASEAN take a position. The Chinese responded by claiming that the Philippines' public outreach violated the mutual agreement to resolve the dispute bilaterally. China responded by replacing the *Zhonggou Haijian* 84 with a Fisheries Law Enforcement Command (FLEC) vessel equipped with several machine guns and light helicopters. Moreover, China informed the Philippines that Chinese warships were deployed to a holding area over the horizon.

On April 23, the Chinese made a gesture of conciliation by announcing that they had withdrawn two of its vessels from the shoal and intended to seek a diplomatic solution. Although Philippine reporting confirmed the withdrawal of these Chinese vessels, the Philippine authorities did not comply but rather deployed a second fishing vessel to the shoal. In addition, the Philippine authorities announced on April 26 that they were seeking direct US involvement. The Chinese reacted immediately by warning that it would respond with its civilian agencies if the situation called for it. Two days later, a Chinese cutter harassed a Philippine vessel at the shoal, and on May 2, four Chinese ships were back in what became known as the Scarborough Shoal standoff.

By mid-June, Philippine President Aquino ordered that all Philippine vessels had to leave the shoal due to an upcoming seasonal typhoon. Once its Coast Guard vessel departed, China began consolidating its control over the shoal by constructing a chain barrier across its narrow mouth, thereby denying the Philippines access to it. At the same time, China's Vice Minister of Foreign Affairs Fu Ying visited Manila and tried to compel the Philippine authorities to keep quiet on the issue and accept Beijing's de facto control over Scarborough Shoal. Today, the West considers the standoff and seizure of the shoal the most tangible disruptive Chinese action in the SCS since 2012. What stands out are the militia's successful actions during the initial stages of the dispute to conduct the preliminary engagement of the shoal. President Xi Jinping conducted a high-profile visit to the Tanmen militia a year after the standoff, indicating the militia's strategic significance.[47]

The militia's involvement in the reclamation of territory is another form of preliminary engagement. The Philippines observed that China employs the militia covertly to construct artificial islands in the Reed Bank area. Between June and September 2023, the Philippines observed numerous militia vessels swarming near Reed Bank. This strategic area in the Philippines EEZ contains many natural resources such as oil and gas. The militia covertly harvests coral in the area and deposits it in a place where China intends to build a new artificial island. The Chinese chose to use coral rather than sand or stones because it doesn't erode with the tides and presents a more sustainable foundation for the artificial islands. The approach indicates the Chinese long-term strategy centered on perseverance. Despite Chinese efforts to conduct these operations undetected, Philippine ISR and underwater operations identified changes in island structures, which they attribute to the militia.

In sum, the militia illustrates an example of tactical units capable of achieving strategic effects, thereby expanding China's control of the SCS. The militias' ambiguous character allows China to effectively manipulate the traditional distinction between military and non-military forces. In doing so, the militia is an effective instrument to be employed below the threshold of armed conflict.

Analysis

Policy

To defend and advance its national interests of state sovereignty, national security, and territorial integrity, China makes several territorial claims in the SCS. It pays particular attention to a chain of islands, defined as the First Island Chain. From an economic perspective, China seeks control of the area to ensure that its merchant vessels can maintain access to the waterways between these islands, connecting mainland China with the Pacific Ocean. By controlling the area, China also ensures access to many natural resources, including fish, oil, and gas. From a military perspective, China wants to increase its military capabilities in the island chain to compensate for the perceived lack of strategic depth between its economically developed coastal areas and the US in case of an armed conflict. To counter this unfavorable balance of power and exploit economic opportunities, gaining control of the SCS is a vital Chinese national interest.

Military strategy

China aims to advance its national interests by employing specific coercive actions as a calculated response to other actors in which costs – mainly relational – serve as a balancing factor. Therefore, the Chinese intend to achieve their objectives gradually over time rather than swiftly, which may require more aggressive actions. To prevent such actions from taking place and be able to operate over extended periods of time in the SCS, China started a sustained irregular warfare campaign in the early 70s. The militia plays a vital role in this campaign and allows China to pursue its goals by exploiting the zone of ambiguity between detection, attribution, and, ultimately, other actors' responses.

Suitability: could the militia conduct the activities with the available resources?

The militia successfully conducted their activities with the available resources. The militia is a very suitable instrument to conduct all three activities that are part of OPE. The effectiveness is mainly the result of the militia's asymmetric advantage created by the fact that it was operationalized from China's civilian fishing fleet. As a result, the Chinese military employs a paramilitary capability that actively swarms throughout the SCS, allowing them to perform OPE-related tasks at short notice. Additionally, the militia has expanded China's ability to collect and communicate information more quickly, thereby providing it with a significant information advantage. The ambiguous character of the militia brings serious challenges to surrounding states when deciding how to respond to activities conducted by civilian fishermen. In other words, by employing the militia, China has created a capability to complete the observe, orient, decide, and act (OODA) loop quicker than other regional states.

Feasibility: did achieving the objectives create the strategic effects that contributed to the desired end?

The militia has achieved most, if not all, of its objectives, thereby creating the strategic effects that contributed to the desired end of gaining control over the SCS. For example, by continuously swarming in the SCS, the militia conducted area assessments and site surveys of potential locations to build artificial islands, including, for instance, near Scarborough Shoal. In the summer of 2023, the militia also successfully conducted surveillance and performed an early warning function, allowing them to communicate critical information about the Philippine resupply mission of the *Sierra Madre* in a timely manner. By doing so, the CCG vessels and regular Navy ships could remain over the horizon until the moment the Philippine mission started. The militia's vessels played a critical role in reporting on the Filipino movements, thereby enabling the Coast Guard vessels and Navy ships to move into an advantageous position at a crucial moment of the mission. Providing the information allowed the Chinese to disrupt the Philippine mission effectively.

In 2012, the militia had successfully seized the Scarborough Shoal, occupying it by using masqueraded fishing vessels. By conducting this small-scale offensive action, the militia allowed the Chinese to conduct follow-on operations and take de facto possession of the Shoal. The subsequent standoff, thus starting in 2012 and continuing to this day, illustrates the difficulty of countering this ambiguous capability, as the Philippine government was reluctant to respond to the events unfolding at the Shoal at that time.

Acceptability: did the strategic effects sought to justify the objectives pursued and consider tangible and intangible costs?

The ambiguous character of the militia prevents other regional states from distinguishing between military and non-military capabilities. In other words, a lack of accurate information complicates decision-making, as states need more procedures and resources to respond effectively to the Chinese fishers acting as paramilitary forces. Although China deliberately violates the normative distinction between combatants and non-combatants, employing its militia hasn't resulted in passing the threshold of armed conflict. Indeed, the militia could generate strategic effects while preventing tactical escalation. The strategic effects, in turn, slowly changed the status quo in favor of China. The *fait accompli* by which China seized Scarborough Shoal is a clear example.

The militia is thus an inherently subversive capability. Subversive activities are "ambiguous because the activity is complex, ill-defined, and hard to observe or attribute."[48] When comparing the activities of the militia conducted over the last four decades with its impact on international relations, it seems that the actions did not come with tangible or intangible costs for China to achieve its objectives. Conversely, if the activities had been conducted by regular Chinese Navy ships, attribution and claiming Chinese hostile intent would not have been difficult and would potentially have led to a military response, ultimately leading to escalation or miscalculation.

In conclusion, the inherent logic of the strategy developed by China proved to be valid. The ends, ways, and means answer the questions of suitability, feasibility, and acceptability. The Chinese learned a valuable lesson from employing the militia. By using ambiguous capabilities when confronting competitors in the SCS, rather than purely military ones, China could render possible responses irrelevant and pursue its objectives at low cost. As a result, China could achieve its strategic objectives and thereby advance its national interests of state sovereignty, national security, and territorial integrity at a low cost.

Risk

What assumptions did the Chinese make in their strategy, and what was the effect of any wrong assumption?

China made several assumptions in its strategy that proved to be correct. For example, the Chinese believed that the militia's shaping activities would allow them to operate below the other actors' redlines. By doing so, the militia achieved its objectives while preventing escalation to armed conflict. Indeed, the employment of the militia illustrates how China correctly assumed that it would be able to control the escalation by hampering regional military action to counter its masqueraded civilian fishing vessels without the risk of using force disproportionally. In other words, China is successful by generating effects in the cognitive dimension (i.e., risk confusion).

Indeed, over the last decades, the Philippine government has remained relatively unresponsive to Chinese actions. In 2023, however, their stance substantially shifted to being more assertive. After two incidents in January 2023, in which China forced Philippine vessels out of Philippine waters, the government expanded its patrols, posture, and exposure. The PCG significantly increased its presence, employed new monitoring capabilities, and expanded air patrols from January to June 2023. The Filipinos also strengthened their posture by intensifying their operations in the SCS. For example, the Philippines showed a more active commitment to the resupply mission at the Second Thomas Shoal and installed five additional navigational buoys at strategic locations surrounding the Spratly Islands to signal the Philippines' claims. Additionally, the Philippine government started the so-called Transparency Initiative to expose Chinese aggression in the SCS. The Philippines released press statements, embedded reporters and politicians on patrols, and allowed video recording. This approach increased news reports, CNN coverage, and international attention. Furthermore, the new Philippine stance was supported by combined multinational patrols, exercises, and statements and by significant US diplomatic efforts with several high-level visits to the region. Together, these activities strengthened and supported the Philippines in its will to resist Chinese coercion but also fueled tensions with China.

Despite a significant shift noticeable in the Philippine stance toward Chinese assertive behavior and recent multinational efforts, a strong and unified response from the Philippine government is still absent, as several senior government officials downplay the significance of China's activities. For example, some politicians lack the political will to stand up against China. Others argue that a whole-of-government approach is required, as the threat of Chinese economic coercion looming in the background could devastate the Philippines. Finally, other politicians are unwilling to respond at all, as they support China. In that regard, the Governor of one of the Philippines' northern provinces stands out, as he deliberately disrupts the Philippine military's training and operations in his area of responsibility.

Did the Chinese incorporate flexibility or adaptability into their strategy? If required, how could they have modified their strategy, and at what cost?

Similar to the Iranian strategy in Lebanon, the Chinese strategy was akin to running a marathon, not a sprint. Regarding strategy, the former needs flexibility over a longer duration, while the latter affords little time for adjustment. As early as the 1970s, Chinese fishing vessels operated in the SCS to support China's national interests. Over time, the Chinese pursued their goals slowly by changing the status quo. What stands out in the sustained campaign is China's ability to advance its territorial claims at a slow pace, thereby encountering only limited reprisals from regional or international actors. This illustrates the flexibility and adaptability of the Chinese strategy, as it centers on the careful calculation that its opponents won't respond to what ultimately will lead to a dramatic shift in regional power structures.

How did other actors react to the Chinese activities?

The international community has widely commented on China's territorial claims in the SCS. The US is the most prominent country that repeatedly has condemned China's claims and coercive actions. In the fall of 2023, various US senior officials, including Vice President Kamala Harris, Secretary of State Anthony Blinken, and Secretary of Defense Loyd Austin, visited several countries in the region, including Indonesia, Vietnam, The Philippines, Vietnam, and Japan. By doing so, the US aims to establish a regional alliance to deter Chinese actions more effectively. Additionally, the US is increasing its military presence in the Philippines, which resulted in several high-profile US military officers visiting the country. Finally, the US publicly reaffirmed its support to the Philippines in case of Chinese aggressive behavior:

> The United States reaffirms an armed attack on Philippine public vessels, aircraft, and armed forces—including those of its Coast Guard in the South China Sea—would invoke US mutual defense commitments under Article IV of the 1951 US Philippines Mutual Defense Treaty.[49]

In 2020, the Australian Government submitted a declaration to the United Nations (UN) stating that "there is no legal basis" for China's territorial and maritime claims in the SCS.[50] By doing so, Australia's stance aligned with the American one. Following the incidents at the Second Thomas Shoal in the summer of 2023, Australia, Japan, Canada, the United Kingdom, and the European Union condemned China's coercive behavior, particularly the dangerous maneuvers several Chinese vessels conducted near PCG vessels.

How did unforeseen factors play a role in the strategy?

During its operations, the militia encountered no difficulties preventing them from achieving their objectives. Thus, no indications of chance or friction playing a role in this strategy were found.

Tactics

Did the militia achieve relative superiority throughout the campaign?

The militia achieved relative superiority throughout the campaign by applying the principle of swarming. The militia is a clear example of a simple tactical unit that can generate strategic effects at a low cost. The Chinese civilian fishing fleet includes more than 2,400 vessels, about 60 percent of which are affiliated with the maritime militia. By comparison, the Philippines registered approximately 900 commercial fishing vessels in 2021. Out of that enormous Chinese fleet, roughly 600 vessels were continuously present in the strategic areas of the SCS. These vessels can sustain their presence at sea for many weeks at a low cost, particularly compared to traditional naval vessels. The militia provides China with superior numbers over the other regional actors, including the Philippines and Vietnam. The numerical advantage does not allow the militia to achieve relative superiority but enables it to achieve a favorable disposition of forces to generate strategic effects.

The militia can realize its favorable position by employing swarming tactics, fulfilling two critical requirements. First, the militia's large fleet consists of many small vessels maneuvering interconnectedly. Besides, they communicate and coordinate with each other securely. By doing so, the militia functions as a "sustainable pulsing force" capable of stealthy coalescing to strike targets unexpectedly from multiple directions, after which it will redisperse to continue swarming in the area while preparing for a new "pulse." The coercive actions that the militia employed in September 2023 at the Second Thomas Shoal are one example of achieving relative superiority through swarming tactics. Second, as a swarming force, the militia conducts strikes and functions as an ISR platform to create and maintain "top sight." The militia persistently performs this early warning function in strategic areas such as the Second Thomas Shoal, Reed Bank, and Scarborough Shoal to maintain an information advantage over the Philippine Navy and Coast Guard. From a strategic perspective, this advantage allowed the militia to achieve relative superiority, that is, a condition where they could decide when, where, and with forces, the follow-on tactical activities were to be conducted.

Having achieved relative superiority through swarming meant that the militia had freedom of movement and action in the SCS, including in the Philippines' EEZ. These advantages indicated that the chances of mission success outweighed the chances of mission failure. Figure 8.2 illustrates this by the gradually inclining curve that reaches further above the Z-axis over time (i.e., relative superiority line), thereby decreasing the area of vulnerability. Of course, the militia operating in the SCS remained vulnerable from the moment relative superiority was achieved. However, they mitigated that vulnerability by manipulating the detection and attribution thresholds, which required strict measures of operational security. For example, the militia turns off the transmitting component of their AIS systems, which prevents them from being tracked and traced. Additionally, their civilian signature allows them to blend in with numerous other civilian vessels. This makes

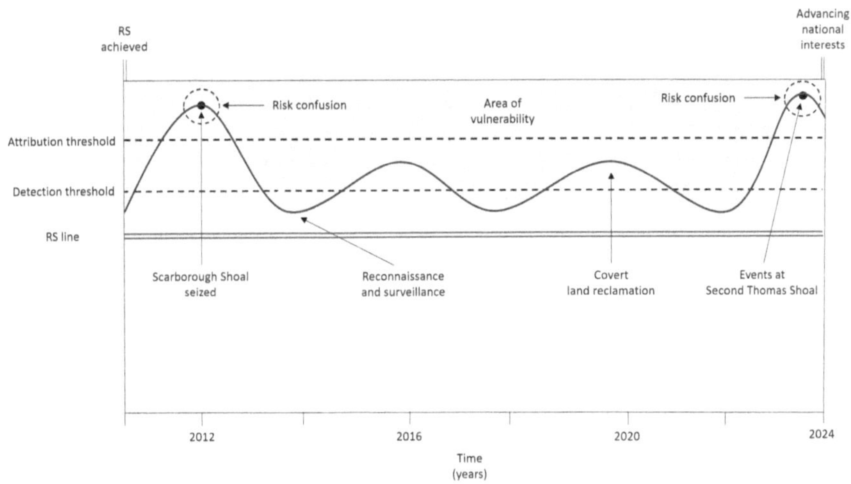

Figure 8.2 Relative superiority graph for China's shaping operations in the SCS.

it difficult for the other actors to distinguish militia vessels from real fishing boats, including the use of ISR capabilities. In other words, the militia hides in plain sight.

If achieved, how did the militia sustain its relative superiority throughout the campaign? If not, why did they fail?

The militia maintained its relative superiority through its profound risk confusion. The militia's ambiguous signature makes it difficult for other actors to counter their activities, as they do not conform neatly to the binary distinction between military and non-military personnel. As a result, regional actors find it difficult to respond appropriately to activities that, at first glance, seem to be conducted by civilian fishers. For example, changes in island structure in the strategic area between Mischief Reef, Rozul Reef, and Sabina Shoal that were identified by Philippine ISR, and its underwater operations could not be attributed to the militia vessels operating in the area at that particular moment. This characteristic reflects the ideas attributed to the Chinese general Sun Tzu, who wrote that a skillful general conceals the disposition of his forces and intent by creating shapes that confuse and delude the enemy.[51]

The Scarborough Shoal standoff is another example of the militia's ambiguous character preventing the Philippine authorities from responding effectively. The case illustrates how the militia had employed its available forces with such skill that they achieved relative superiority over the Philippine Navy and Coast Guard at the decisive point. Indeed, by aligning the operational factors of time and space, the militia could influence the event's outcome considerably. The civilian signature of the Chinese vessels, who were already inside the Shoal before the Filipinos arrived, had a decisively disruptive effect on the government's strategic risk calculations. The *fait accompli* by which the militia had seized the Shoal illustrates that the risk of deciding between action or inaction appeared just as tremendous and unattractive.

Conclusion

The militia forms the tip of the spear of China's sustained irregular warfare campaign in the SCS. The maritime militia advances China's national interests in the area by achieving its objectives without using armed force and, importantly before China's competitors can deliver an effective response. Sun Tzu asserted that "to win one hundred victories in one hundred battles is not the acme of skill. To subdue the enemy without fighting is the acme of skill."[52] The Chinese can do so throughout the campaign because of the militia's ambiguous character, which allows them to manipulate the detection and attribution thresholds, thereby rendering responses irrelevant.

The protracted campaign illustrates that China's political leaders and military commanders have demonstrated a high-level of strategic sense, as they well understood converting tactical performance into strategic effect. The Chinese campaign thus fulfilled the strategic promise of special operations by carefully aligning policy, strategy, and tactics.

Notes

1 State Council Information Office of the People's Republic of China, *China's Peaceful Development*, September 6, 2011.
2 Feng Zhaokui, "What Are China's Core Interests?" *China-US Focus*, October 21, 2014.
3 Ibid.
4 International Maritime Organization, "United Nations Convention on the Law of the Sea."
5 Shahar Hameiri and Lee Jones, *Fractured China: How State Transformation Is Shaping China's Rise* (New York: Cambridge University Press, 2021), 85.
6 Permanent Court of Arbitration, The South China Sea Arbitration (The Republic of Philippines v. The People's Republic of China), July 12, 2016.
7 Anthony H. Cordesman, Arleigh A. Burke, and Max Molot, "The Critical Role of Chinese Trade in the South China Sea," *Center for Strategic and International Studies*, October 1, 2019.
8 Ibid.
9 Liu Huaqing, *Selected Military Writings of Liu Huaqing* (Beijing: People's Liberation Army Press, 2008).
10 Liu Baoyin and Yang Ziaomei, *Island Chain Surrounding China-Oceanic Geography, Military Location, and Information Systems* (Beijing: Haiyang Press, 2003).
11 Ben Dolven, Susan V. Lawrence, and Ronald O'Rourke, "China Primer: South China Sea Disputes" (Congressional Research Service (CRS)), December 22, 2020.
12 Yu Kaijin, Li Guangsuo, and Cao Yongheng, "An Analysis of the Island Chain," *Ship and Boat* 5 (2006).
13 Hu Bo, "On China's Important Maritime Interests," *Asia-Pacific Security and Maritime Research* 3 (2015).
14 Ronald O'Rourke, "US-China Strategic Competition in South and East China Seas: Background and Issues for Congress," *Congressional Research Service,* February 8, 2023.
15 Stephen Biddle and Ivan Oelrich, "Future Warfare in the Western Pacific: Chinese Anti Access/Area Denial, US AirSea Battle, and Command of the Commons in East Asia," *International Security* 41, no. 1 (2016).
16 O'Rourke, "US-China Strategic Competition," 2–3.
17 Marc Lanteigne, *Chinese Foreign Policy: An Introduction*, 4th ed. (London: Routledge, 2019).

18 Stein Tonnesson, "The Paracels: The Other South China Sea Dispute," *Peace Research Institute Oslo*, July 2001.
19 Lanteigne, *Chinese Foreign Policy*.
20 Ketian Zhang, "Cautious Bully: Reputation, Resolve, and Beijing's Use of Coercion in the South China Sea," *International Security* 44, no. 1 (2019).
21 M. Taylor Fravel, "Threading the Needle: The South China Sea Disputes and US-China Relations," *SSRN Scholarly Paper* (Rochester, NY, July 8, 2016).
22 Hameiri and Jones, *Fractured China*.
23 Fravel, "Threading the Needle."
24 Zoe Haver, "China Maritime Report No. 12: Sansha City in China's South China Sea Strategy: Building a System of Administrative Control," *CMSI China Maritime Reports*, January 1, 2021.
25 Fravel, "Threading the Needle."
26 Ibid.
27 Michael J. Green, *Countering Coercion in Maritime Asia: The Theory and Practice of Gray Zone Deterrence*, CSIS Reports (Washington, DC: Lanham, MD: Center for Strategic & International Studies; Rowman & Littlefield, 2017).
28 Fravel, "Threading the Needle."
29 "China Lands First Bomber on South China Sea Island," Asia Maritime Transparency Initiative, accessed December 13, 2022; Amanda Macias, "China Quietly Installed Missile Systems on Strategic Spratly Islands in Hotly Contested South China Sea," *CNBC*, May 2, 2018.
30 O'Rourke, "US-China Strategic Competition."
31 Jane Perlez, "Rodrigo Duterte and Xi Jinping Agree to Reopen South China Sea Talks," *The New York Times*, October 20, 2016.
32 John Grady, "SECNAV, Navy Maritime Intel Director Give Warnings About Illegal Chinese Fishing," *USNI News*, March 6, 2023.
33 Conor Kennedy and Andrew Erickson, "China Maritime Report No. 1: China's Third Sea Force, The People's Armed Forces Maritime Militia: Tethered to the PLA," *CMSI China Maritime Reports* 1 (2017).
34 "国防后备力量" [National Defense Reserve Force], Jinhu County People's Government Double-Support Office, 20 October 2016.
35 Kennedy and Erickson, "China Maritime Report."
36 Shinji Yamaguchi, "Strategies of China's Maritime Actors in the South China Sea," *China Perspectives; Wanchai* 3 (2016).
37 "三沙市推动军警民联防机制 构建三线海上维权格局" [Sansha City Advances its Joint Military, MLE, Militia Defense System—Constructs a Three-line Maritime Rights Protection Layout], *China News Online*, 22 November 2014; Interview with PLA Major General Liu Lianhua: "军队代表: 应从战略层面构建军警民海防体系" [Military Representative: [We] Should Strategically Construct a Military, Police, Militia Maritime Defense System], China News Online, 15 March 2013.
38 Kennedy and Erickson, "China Maritime Report."
39 Abhijit Singh, "Deciphering Grey-Zone Operations in Maritime-Asia," ORF (blog).
40 "破解海上民兵建设难题" [Resolving Issues in Maritime Militia Construction], National Defense News, 28 July 2016; 张践 [Zhang Jian], "围绕'六化'抓建 推动海上民兵转型" [Advance Transformation of the Maritime Militia Centered on Six Changes], National Defense 10 (2015).
41 Department of Defense, Joint Publication 3-13, *Joint Doctrine for Information Operations* (Washington D.C: Department of Defense, 1998).
42 Linda Robinson, "Plan of Attack, The Pentagon Has a New Strategy for Taking on Terrorists-and Taking Them Down." *US News and World Report*, 01 August 05, 3.
43 Michael T. Kenny, Leveraging Operational Preparation of the Environment in the GWOT 3, US Army Command and General Staff College, May 25, 2006.

44 Michael S. Repass, Combating Terrorism with Preparation of the Battlespace, US Army War College, April 7, 2003.
45 John Arquilla and David Ronfeldt, "Swarming and the Future of Conflict" (Santa Monica, CA: RAND Corporation, 2000).
46 Thai News Service Group, "China/Philippines: China Seeks Preservation of Over-All Friendly Relations with Philippines as Tension over Scarborough Shoal Ebbs Monetarily," *Asia News Monitor*, April 12, 2012.
47 Conor Kennedy and Andrew Erickson, "Model Maritime Militia: Tanmen's Leading Role in the April 2012 Scarborough Shoal Incident," CIMSEC, April 21, 2016.
48 Thomas C. Schelling, *Arms and Influence* (New Haven, CT: Yale University Press, 2008), 77.
49 US Department of State, "US Support for the Philippines in the South China Sea," *Press Releases*, August 5, 2023.
50 Elias Visontay, "Australia Declares 'There is No Legal Basis' to Beijing's Claims in South China Sea," *The Guardian*, July 24, 2020.
51 Sun Tzu, in *The Art of War*, trans. Samuel B. Griffith (Oxford University Press, 1963).
52 Sun Tzu, *The Art of War*.

Bibliography

Arquilla, John, and David Ronfeldt. *Swarming and the Future of Conflict*. Santa Monica, (CA: RAND Corporation, 2000), https://www.rand.org/pubs/documented_briefings/DB311.html.

Baoyin, Liu, and Yang Ziaomei. *Island Chain Surrounding China-Oceanic Geography, Military Location, and Information Systems* (Beijing: Haiyang Press, 2003).

Biddle, Stephen, and Ivan Oelrich. "Future Warfare in the Western Pacific: Chinese Anti Access/Area Denial, US AirSea Battle, and Command of the Commons in East Asia," *International Security* 41, no. 1 (2016).

Bo, Hu. "On China's Important Maritime Interests," *Asia-Pacific Security and Maritime Research* 3 (2015).

"China Lands First Bomber on South China Sea Island," Asia Maritime Transparency Initiative, May 18, 2018, https://amti.csis.org/china-lands-first-bomber-south-china-sea-island/.

Cordesman, Anthony H., Arleigh A. Burke, and Max Molot. "The Critical Role of Chinese Trade in the South China Sea" Center for Strategic and International Studies, October 1, 2019, https://www.jstor.org/stable/resrep22586.30.

Department of Defense. *Joint Publication 3-13, Joint Doctrine for Information Operations* (Washington D.C: Department of Defense, 1998).

Dolven, Ben, Susan V. Lawrence, and Ronald O'Rourke. "China Primer: South China Sea Disputes" (Congressional Research Service (CRS)), December 22, 2020, https://crsreports.congress.gov/product/pdf/IF/IF10607/4.

Fravel, M. Taylor. "Threading the Needle: The South China Sea Disputes and US-China Relations," SSRN Scholarly Paper (Rochester, NY, July 8, 2016), https://doi.org/10.2139/ssrn.2807181.

Grady, Jonh. "SECNAV, Navy Maritime Intel Director Give Warnings About Illegal Chinese Fishing," *USNI News*, March 6, 2023, https://news.usni.org/2023/03/06/secnav-navy-maritime-intel-director-give-warnings-about-illegal-chinese-fishing.

Green, Michael J. *Countering Coercion in Maritime Asia: The Theory and Practice of Gray Zone Deterrence*, CSIS Reports (Washington, DC: Lanham, MD: Center for Strategic & International Studies, Rowman & Littlefield, 2017).

Haver, Zoe. "China Maritime Report No. 12: Sansha City in China's South China Sea Strategy: Building a System of Administrative Control," *CMSI China Maritime Reports*, January 1, 2021, https://digital-commons.usnwc.edu/cmsi-maritime-reports/12.

Hameiri, Shahar, and Lee Jones. *Fractured China: How State Transformation Is Shaping China's Rise* (New York: Cambridge University Press, 2021).
Huaqing, Liu. *Selected Military Writings of Liu Huaqing* (Beijing: People's Liberation Army Press, 2008).
International Maritime Organization. "United Nations Convention on the Law of the Sea," https://www.imo.org/en/OurWork/Legal/Pages/UnitedNationsConventionOnThe-LawOfTheSea.aspx.
Interview with PLA Major General Liu Lianhua: "军队代表: 应从战略层面构建军警民海防体系" [Military Representative: [We] Should Strategically Construct a Military, Police, Militia Maritime Defense System], China News Online, 15 March 2013, http://www.chinacourt.org/article/detail/2013/03/id/922031.shtml.
Kaijin, Yu, Li Guangsuo, and Cao Yongheng. "An Analysis of the Island Chain," *Ship and Boat* 5 (2006).
Kenny, Michael T. Leveraging Operational Preparation of the Environment in the GWOT 3, US Army Command and General Staff College, May 25, 2006), https://man.fas.org/eprint/kenny.pdf.
Kennedy, Conor, and Andrew Erickson. "Model maritime militia: Tanmen's leading role in the April 2012 Scarborough shoal incident," CIMSEC, April 21, 2016, https://cimsec.org/model-maritime-militia-tanmens-leading-role-april-2012-scarborough-shoal-incident/.
Kennedy, Conor, and Andrew Erickson. "China Maritime Report No. 1: China's Third Sea Force, The People's Armed Forces Maritime Militia: Tethered to the PLA," *CMSI China Maritime Reports* 1 (2017), https://digital-commons.usnwc.edu/cmsi-maritime-reports/1.
Lanteigne, Marc. *Chinese Foreign Policy: An Introduction* (London: Routledge, 2019).
Macias, Amanda. "China Quietly Installed Missile Systems on Strategic Spratly Islands in Hotly Contested South China Sea," CNBC, May 2, 2018, https://www.cnbc.com/2018/05/02/china-added-missile-systems-on-spratly-islands-in-south-china-sea.html.
"国防后备力量" [National Defense Reserve Force]. Jinhu County People's Government Double-Support Office, 20 October 2016, http://www.jinhu.gov.cn/art/2016/10/20/art_96_19469.html. (Not accessible as of [30/07/2024])
O'Rourke, Ronald. "US-China Strategic Competition in South and East China Seas: Background and Issues for Congress," Congressional Research Service, February 8, 2023, https://crsreports.congress.gov/product/pdf/R/R42784/141.
Perlez, Jane. "Rodrigo Duterte and Xi Jinping Agree to Reopen South China Sea Talks," *The New York Times*, October 20, 2016, https://www.nytimes.com/2016/10/21/world/asia/rodrigo-duterte-philippines-china-xi-jinping.html.
Permanent Court of Arbitration. The South China Sea Arbitration (The Republic of Philippines v. The People's Republic of China), July 12, 2016, https://pca-cpa.org/en/cases/7/.
Repass, Michael S. Combating Terrorism with Preparation of the Battlespace, US Army War College, April 7, 2003, https://apps.dtic.mil/sti/tr/pdf/ADA414205.pdf.
"破解海上民兵建设难题" [Resolving Issues in Maritime Militia Construction], National Defense News, 28 July 2016, www.81.cn/gfbmap/content/2016-07/28/content_151895.htm.
Robinson, Linda, "Plan of Attack, The Pentagon Has a New Strategy for Taking on Terrorists-and Taking Them Down." US News and World Report, August 1, 2005, https://stationhypo.com/2023/05/26/demystifying-the-title-10-title-50-debate-distinguishing-military-operations-intelligence-activities-covert-action-part-3-of-4/.
"三沙市推动军警民联防机制 构建三线海上维权格局" [Sansha City Advances its Joint Military, MLE, Militia Defense System—Constructs a Three-line Maritime Rights Protection Layout], *China News Online*, 22 November 2014, http://www.chinanews.com/gn/2014/11-21/6803776.shtml.
Schelling, Thomas. *Arms and Influence* (New Haven, CT: Yale University Press, 2008).
Singh, Abhijit. "Deciphering Grey-Zone Operations in Maritime-Asia," ORF (blog). https://www.orfonline.org/research/42978 deciphering-grey-zone-operations-in-maritime-asia/.

State Council Information Office of the People's Republic of China. *China's Peaceful Development*, September 6, 2011. https://english.www.gov.cn/archive/white_paper/2014/09/09/content_281474986284646.htm

Thai News Service Group. "China/Philippines: China Seeks Preservation of Over-All Friendly Relations with Philippines as Tension over Scarborough Shoal Ebbs Monetarily," *Asia News Monitor*, April 12, 2012.

Tonnesson, Stein. The Paracels: The Other South China Sea Dispute – Peace Research Institute Oslo," Peace Research Institute Oslo, July 2001, https://www.prio.org/publications/2270.

Tzu, Sun. *The Art of War*. Translated by Samuel B. Griffith (Oxford University Press, 1963).

US Department of State. "US Support for the Philippines in the South China Sea," Press Releases, August 5, 2023, https://www.state.gov/u-s-support-for-the-philippines-in-the-south-china-sea-5/.

Visontay, Elias. "Australia declares 'there is no legal basis to Beijing's claims in South China Sea," The Guardian, July 24, 2020, https://www.theguardian.com/australia-news/2020/jul/25/australia-declares-there-is-no-legal-basis-to-beijings-claims-in-south-china-sea.

Yamaguchi, Shinji. "Strategies of China's Maritime Actors in the South China Sea," China Perspectives; *Wanchai* 3 (2016).

张践 [Zhang Jian]."围绕'六化'抓建 推动海上民兵转型" [Advance Transformation of the Maritime Militia Centered on Six Changes], National Defense 10 (2015), http://kns55.en.eastview.com/kcms/detail/detail.aspx?recid=&FileName=GUOF201510009&DbName=CJFD2015&DbCode=CJFD.

Zhang, Ketian. "Cautious Bully: Reputation, Resolve, and Beijing's Use of Coercion in the South China Sea," *International Security* 44, no. 1 (2019).

Zhaokui, Feng. "What Are China's Core Interests?" China-US Focus, October 21, 2014, https://www.chinausfocus.com/foreign-policy/what-are-chinas-core-interests-2.

Part IV

Conclusions

9 Fulfilling the Strategic Promise of Special Operations

Introduction

Special operations are a miraculous tool to advance national interests, as their positive disproportionate return on investment goes against the most fundamental principles of military strategy. One of history's most influential military theorists, Carl von Clausewitz, argued that defense is the more potent form of warfare.[1] He further stated that the first principle in strategy is to employ the greatest number of forces at the decisive point in battle.[2] Yet, special operations, often conducted with a limited number of forces operating deep in enemy territory and with little support, prove to be an effective tool to exploit specific vulnerabilities in enemy defenses, which typically have superior numbers, and achieve strategic objectives. The ability to overcome these significant challenges and achieve major foreign policy goals makes special operations a strategic instrument. Against this background, the main research question to be answered in this work is: "How do the personnel involved fulfill the strategic promise of special operations, given their position within the larger framework of foreign policy?"

The case study findings revealed that fulfilling the strategic promise of special operations depends on how (i.e., tactics and operations) the leadership of the state employs the instruments of power (i.e., means) to achieve specific objectives and, importantly, at what cost. From that perspective, policymakers orchestrate the employment of the instruments of power, which serve as the bases for military commanders or others responsible for exercising command and control over special operations, to plan, prepare, and execute these endeavors. The strategic design and tactical execution are thus an integrated whole that should ideally operate synergistically.

In exceptional cases, the personnel involved will be able to bridge the potential gap between the political objectives, and the appropriate tactics by creating a joint venture, allowing a successful outcome of a special operation. Despite the clear distinction between responsibilities, exploiting the full potential of special operations will demand that policymakers are fully aware of what is achievable by employing the instruments of power. Military commanders must be mindful of the political objective, as well as experts using the capabilities at their disposal and, most importantly, fully aware of their limitations. Thus, special operations

DOI: 10.4324/9781003365112-13

depend on effective civil-military relationships wherein policymakers and military commanders are aligned in the operation's purpose and complementary in actions.

These outstanding achievements, displayed in several of the presented case studies, while lacking in others, illustrate the need for a new conception of special operations, namely, to consider them as an integrating function. As such, policymakers continue to employ assets derived from all instruments of power in pursuit of their objectives. At the same time, the personnel responsible for the tactical execution, the military commanders included, will use these assets across all domains to achieve relative superiority over an opponent, a condition that will result in strategic effects.

Through this grand strategic perspective, the people involved in special operations will be able to select and integrate policies, strategies, and tactics, feasibly and suitably, as well as at an acceptable cost. This is how they display a strategic sense and, ultimately, utilize the full potential of special operations.

Policies

A key finding from the case studies is that special operations are distinct from conventional military operations, as they are a grand strategic tool that presents unique options for policymakers to defend, and advance national interests. The specific objectives of special operations vary between classical military ones, such as described by Clausewitz, and non-military ones.[3] The latter may include gaining political influence, protecting economic interests, and neutralizing specific threats, including nuclear ones. Thus, special operations are a unique strategic instrument that policymakers can employ to attain specific goals in peacetime, as well as in war.

From a strictly military perspective, special operations are unique ways of employing means to achieve classical military objectives. According to Clausewitz, the purpose of the military instrument of power is threefold: destroy an enemy's armed forces (i.e., forcing them into a situation from which they can no longer sustain their fight), occupy the enemy's country, and break their will to continue fighting.[4]

The failed Russian raid on Hostomel Airport, described in Chapter 3, is the only case study that exemplifies how special operations may support a broader politico-military campaign to achieve these three objectives. The Russian Annexation of Crimea described in Chapter 6 includes one of Clausewitz's objectives: seizing territory. What stood out in the latter case was that the Russians achieved this objective by directing their primary efforts to deceive the Ukrainians, rather than fighting them, and breaking their will.

The objectives of the six special operations described, and on which their tactical action was based, varied just as much as their political goals. From a grand strategic perspective, special operations are unique ways of employing means to achieve a broad range of objectives, including non-military ones. The advantage obtained by destroying the enemy is always immense. However, apart from having a certain price, it will usually involve serious risks. To avoid these risks and costs, policymakers may select other options to pursue their political objectives.[5]

Iran's proxy war in Lebanon, described in Chapter 7, originally had multiple objectives. First, the Iranians wanted to export the Islamic Revolution to Lebanon, and maintain power and influence over Hezbollah and, by extension, Lebanese politics. Over time, they also had their sights on extrapolating the Hezbollah model to other countries in the Middle East to increase their presence and influence throughout the entire region, while simultaneously decreasing Western influence. The Chinese operations in the South China Sea (SCS), described in Chapter 8, also had multiple objectives. Fundamentally, the Chinese wanted to seize territory. They did so by preventing their opponents from coming into action at the decisive moment rather than using force. However, the objectives of the campaign were much broader. First, China wanted to increase its military capabilities in the first island chain to compensate for the perceived lack of strategic depth between its economically developed coastal areas and the United States (US), in case of an armed conflict. Next, the Chinese also had an economic goal: seeking control of the area to ensure its merchant vessels could maintain unlimited access to the waterways between these islands, connecting mainland China with the Pacific Ocean. Finally, by controlling the area, China consolidated its access to many natural resources, including fish, gas, and oil. These multiple objectives reflect the broad political purposes associated with these protracted, and still ongoing Iranian and Chinese operations in Lebanon and the SCS, respectively.

The objectives of Operations Orchard and Olympic Games described in Chapters 4 and 5 do not correspond with the three objectives described by Clausewitz. During Operation Orchard, the Israelis targeted a specific piece of Syrian infrastructure (i.e., the nuclear installations) that posed a significant threat to their national security. From that perspective, the operation's objective was to disarm the Syrians rather than destroy them. Operation Olympic Games provides similar insights. Instead of destroying the Natanz facility, the attackers pursued their objective of delaying uranium enrichment without the classical way of using force. Although of strategic importance, the objectives of both operations were more limited than those described by Clausewitz, and both reflected political goals seriously differing from those traditionally associated with the use of force.

Strategies

The findings from the case studies clearly show that special operations are typical examples of grand strategic tools. As such, special operations are all-inclusive: They represent "all assets of a security community for the purposes of policy" – as defined by Gray and Johnson.[6] From that perspective, special operations are not exclusively composed of military capabilities (i.e., means), but are much more versatile, as their capabilities may be derived from all the instruments of national power. For example, Operations Orchard and Olympic Games illustrate the crucial role of intelligence collection during the preparation phases to achieve an information and decision advantage that proved vital for the execution phases. The annexation of Crimea illustrates the importance of combining the military instrument with the diplomatic, information, and intelligence instruments of power. Iran's proxy war in

Lebanon and the Chinese campaign in the SCS are even broader in scope and are composed of capabilities derived from almost all instruments of national power. This integrated view of special operations mirrors the spider web pattern: policymakers continue to add capabilities to a special operation to pursue their objectives, just as the spider continues to expand its web until it catches its prey. This metaphor reflects the creativity required to identify the desired instruments of power to meet the challenges and exploit the opportunities of the strategic environment.

The expansion of capabilities reflects the variety of personnel that are involved in special operations. The raid on Hostomel Airport is an example of a special operation conducted exclusively by Russian elite military forces. A mixture of conventional military forces, SOF, and intelligence personnel conducted the Annexation of Crimea and Operation Orchard. Operation Olympic Games was conducted by a variety of intelligence personnel, including those specialized in Human Intelligence (HUMINT), Imagery Intelligence (IMINT), Signals Intelligence (SIGINT), and Computer Network Operations (CNO). What stands out is that multiple countries deployed these capabilities. Iran's proxy war in Lebanon also included a variety of personnel, including, among others, diplomats, Islamic Revolutionary Guard Corps (IRGC) operatives, and religious scholars. Finally, the Chinese maritime militia, a paramilitary capability, played a crucial role in the operations in the SCS. These findings indicate that the potential ways to meet the challenges and exploit the opportunities of the strategic environment determine what type of personnel will be employed to conduct special operations, pursuing their specific objectives. The forces that conduct special operations, therefore, differ in varying degrees from conventional military forces and are not strictly military (i.e., military forces, paramilitary forces, or intelligence personnel, among others), as the selection of the personnel that will be involved, depends on the specific strategic context in which the operation is planned.

Tactics

Direct special operations

Achieving and sustaining relative superiority during direct special operations remains vital for mission success. During the raid on Hostomel Airport, the Russians successfully applied the principle of surprise, which, from the airport defenders' perspective, allowed the strike force to arrive at its target undetected. At that point, the Russians had achieved relative superiority. They then applied the principle of speed and purpose, which allowed them to sustain their relative superiority, and seize the airport two hours later. The case, thus, illustrates one in which the Russians achieved this decisive condition through superior tactics.

Apart from gaining relative superiority through tactics, the emergence of new domains of military operations has greatly expanded the potential ways of achieving relative superiority. During Operation Orchard, the Israelis successfully manipulated the Syrian Integrated Air Defense System (IADS) by using the electromagnetic spectrum (EMS). By doing so, the Israelis took the Syrians entirely by surprise, which enabled them to achieve relative superiority from the onset of the operation. This condition allowed the Israeli strike force to enter Syrian airspace undetected, destroy its target, and safely return to base within a four-hour window. The operation, therefore, illustrates how advanced technologies allow for new ways of applying the principle of surprise (i.e., surprise by deception), which contributes to achieving and sustaining relative superiority. The same applies to Operation Olympic Games, where the attackers used cyberspace to infiltrate the Stuxnet malware into the Local Area Network (LAN) of the Natanz site, and take complete control of the centrifuges while remaining undetected. At that point, the attackers had achieved relative superiority. Similar to the Electronic Attack (EA) that targeted the Syrian IADS, Stuxnet used a spoofing attack to deceive the Iranian engineers monitoring the Supervisory Control and Data Acquisition (SCADA) systems. This action allowed the attackers to sustain their relative superiority. Accordingly, achieving relative superiority through the principles of surprise, speed, and purpose, as described by McRaven, remains just as relevant as it was when he proposed his six principles. The potential ways of achieving this condition, however, have expanded: next to superior tactics, advanced technologies connected to cyberspace and the EMS have become a second category through which relative superiority can be achieved.

In all three cases, the strike forces achieved relative superiority over their opponents prior to the engagement. In the case of the raid on Hostomel Airport, the helicopter insertion, of which the airport's defenders remained unaware, allowed the Russians to achieve relative superiority prior to the engagement. In Operations Orchard and Olympic Games, the ability to achieve relative superiority prior to the engagement was an integral part of the planning and preparation phases. Both operations were planned and prepared over several years, leading to an information advantage and, by extension, a decision advantage in selecting and developing the most suitable attack method.

The supposition that regaining relative superiority is difficult. Once it's lost, remains valid today. The Russians lost their relative superiority by failing to remove the heavy equipment from the runway, which led to the cancellation of the TALO. This failure illustrates poor Russian planning and preparation, as they could have expected such defensive actions to occur at an airport vulnerable to air assault and TALO. Once the Russians had lost their relative superiority, they could not regain it. In the case of Operation Olympic Games, the attackers lost their relative superiority once Stuxnet was unleashed on the internet. The Iranians quickly learned about the malware, and shut the installation down. This meant that detection immediately led to the loss of relative superiority. Importantly, detection also meant the end of the mission, as the site's systems received the required security patches shortly afterward.

A final finding from the case studies of direct special operations is that the execution phase of these operations could have a protracted character. The execution phase of a direct special operation is typically short in duration, ranging between minutes and hours, depending on the precision-strike capability. The findings from Chapter 5 illustrate that the possibility of conducting such operations through cyberspace dramatically changes the notion of time, as the domain allows for a protracted execution phase.

In sum, the ability to conduct special operations in cyberspace illustrates a paradigm shift, as the new domain requires significant changes in how the people involved in direct special operations think about the planning, preparation, and execution of these endeavors that complement and partially replace the prior paradigm of the classic ones conducted in the air, land, and sea domains.

Indirect special operations

The longer it takes to achieve relative superiority, the greater the risk of detection and attribution. The element of risk is known as the area of vulnerability. Reflecting on the typical short execution phase of direct special operations, relative superiority may be quickly achieved directly before achieving the objective. Such missions are, therefore, characterized by a relatively brief period of vulnerability. By contrast, the typical long duration of indirect special operations indicates that achieving relative superiority may require significant time, if ever really achieved. Consequently, these missions are characterized by a more extended period of vulnerability. Furthermore, given that McRaven's work focused on direct special operations, we should also ask ourselves to what extent the six principles he identified apply to this other type of special operations.

The case study findings illustrate that achieving and sustaining relative superiority during indirect special operations is also a vital condition for mission success but is achieved differently and requires applying other principles of war. Similar to the three case studies that focused on direct special operations, the annexation of Crimea presents a case in which the Russians achieved relative superiority prior to the engagement (i.e., the annexation) by applying the principle of surprise (i.e., clandestine preparations). The Russians sustained their relative superiority by applying the principles of speed (i.e., moving up the date of the referendum), surprise (i.e., surprise by deception), security, and, importantly, legitimacy (i.e., exploiting *bravatso*). Thus, the case reflects one in which relative superiority was achieved through superior tactics. The case study on Iran's proxy war in Lebanon illustrates how the IRGC also achieved relative superiority by applying the principle of legitimacy to gain popular support in the Shia-dominated areas. The IRGC sustained its relative superiority by applying the principle of perseverance, like in a marathon. Importantly, the IRGC turned out to be so effective because of its in-depth networks with a variety of regional partners, which allowed them freedom of movement throughout critical parts of the region. The IRGC, thus, achieved and sustained its relative superiority organizationally, i.e., through network building. Finally, the case study on China's campaign in the SCS demonstrates that the maritime militia achieved relative superiority by applying the principle of maneuver, i.e., employing the concept of swarming at the operational and tactical levels, as well as the principle of security. The maritime militia sustained its relative superiority through its ambiguous signature, which led to risk confusion. This case study presents the final two categories through which relative superiority can be achieved and sustained, namely doctrinally (i.e., the swarming concept) and cognitively (i.e., risk confusion).

In sum, achieving relative superiority as a critical condition for mission success is considered a constant factor for direct and indirect special operations. The potential ways of achieving this condition were aligned with the ability to conduct special operations across the five domains of military operations and the EMS to meet the challenges and opportunities of the strategic environment. These include tactical, technical, organizational, doctrinal, and cognitive ways. The correct calculation of time, space, and force has become the most crucial factor in achieving relative superiority across multiple domains of military operations. Moreover, extrapolating relative superiority theory led to identifying the principles of war that can be controlled and that can influence the achievement of relative superiority during indirect special operations. Next to the six principles identified by McRaven, the case study findings on indirect special operations revealed that legitimacy and perseverance are vital principles for such endeavors. These insights, derived from six distinct cases, inform the personnel involved in special operations on what a decisive advantage across domains looks like, which principles of war come into play, and how these insights can be applied to the planning, preparation, and execution of special operations.

Closure

The case study findings offer a broad perspective on the strategic utility of special operations. This perspective includes the application of the instruments of power, the employment of capabilities across all domains, ways to achieve and sustain relative superiority, and determining the principles of war that can be controlled and can influence tactical behavior. Accordingly, the planning, preparation, and execution of special operations, including the major mission command activities of the operations process, must be integrated across time, space, and force. The integration of capabilities should serve the operation's requirements, rather than being a goal itself. During the planning, preparation, and execution of special operations, therefore, detailed attention must be given to identifying and visualizing the potential integration of these capabilities since each imposes unique geographical and time considerations regarding their employment. This reasoning explains the new conception of special operations, namely, to consider them as an integration function.

Viewing special operations as an integrating function explains how capabilities across the individual domains relate to each other to achieve integration. The network, integrating the various capabilities employed across the individual domains, supports the strength, type, and scope of the relationships between the various domains. Precisely, this integration will generate cross-dimensional effects effectively. As an integrating function, the employment of assets across domains will give rise to emergent physical, virtual, or cognitive capabilities that could not be achieved by an individual asset, but can emerge from integrating two or more assets that collectively generate cross-dimensional effects. Such a system's perspective will facilitate bilateral consultation as well as a mutual understanding of strategy development and tactical behavior. Thus, viewing special operations as an integrating function enables the people involved in these endeavors to display a strategic sense. This is how they fulfill the strategic promise of special operations.

Notes

1 Carl von Clausewitz, *On War*, ed. and trans. Michael Howard, Peter Paret, and Bernard Brodie (Princeton, N.J.: Princeton University Press).
2 Ibid.
3 Ibid.
4 Ibid.
5 Ibid.
6 Colin S. Gray and Jeannie L. Johnson, "The Practice of Strategy," in *Strategy in the Contemporary World: An Introduction to Strategic Studies*, 6th ed., ed. John Baylis, James Wirtz, Eliot Cohen, and Colin S. Gray, (New York: Oxford University Press, 2019).

Bibliography

Clausewitz von, Carl, *On War*. eds. Michael Howard and Peter Paret, (New York: Princeton University Press, 1984).
Gray, Colin S., and Jeannie L. Johnson, "The Practice of Strategy," in *Strategy in the Contemporary World: An Introduction to Strategic Studies*, 6th ed., eds. John Baylis, James Wirtz, Eliot Cohen, Colin S. Gray (New York: Oxford University Press, 2019, 6th Edition).

Index

Note: Page numbers in *italics* refer to figures.